C. Farewell
Golden B.
D'Urville I.
TASMAN BAY
Motueka
Ur. Motueka
Nelson
Mapua
Richmond
Picton
Awatere R.
Blenheim
Seddon
Richmond Brook
C. Campbell
Westport
C. Foulwind
Buller R.
Tapuaenuku 9465'
Clarence R.
Reefton
Lewis Pass
Kaikoura
Runanga
Greymouth
Brunner
Hanmer
Hokitika
Cheviot
Mendip H. Stn.
Waiau R.
Hurunui R.
Arthur's P.
Whataroa R.
Oxford
Rangiora
Kaiapoi
Waimakariri R.
Franz Josef Glacier
Fox Glacier
CHRISTCHURCH
Rakaia
Southern Alps
Mt. Tasman 11475'
Mt. COOK 12349'
Methven
BANKS PENINSULA
Carters Mill
Hermitage
Akaroa
Haast
L. Tekapo
Ashburton
Jackson B.
Mt. Brewster 8264'
Fairlie
Geraldine
Turnbull R.
Haast R.
L. Pukaki
Temuka
L. Ohau
Mt. Aspiring 9957'
L. Benmore
TIMARU
Milford Sd.
L. Hawea
L. Wanaka
Waimate
Homer Tunnel
Waitaki R.
Queenstown
Arrowtown
Oamaru
Remarkables Station
Wakatipu
L. Te Anau
Alexandra
Doubtful Sd.
Palmerston
L. Manapouri
Mosgiel
Port Chalmers
DUNEDIN
Dusky Sd.
Oreti R.
Dipton
Mataura R.
L. Hauroko
Gore
Milton
Balclutha
Winton
Mataura
Kaitangata
Wyndham
Clutha R.
Riverton
INVERCARGILL
Bluff
Foveaux Str.
Oban
STEWART ISLAND

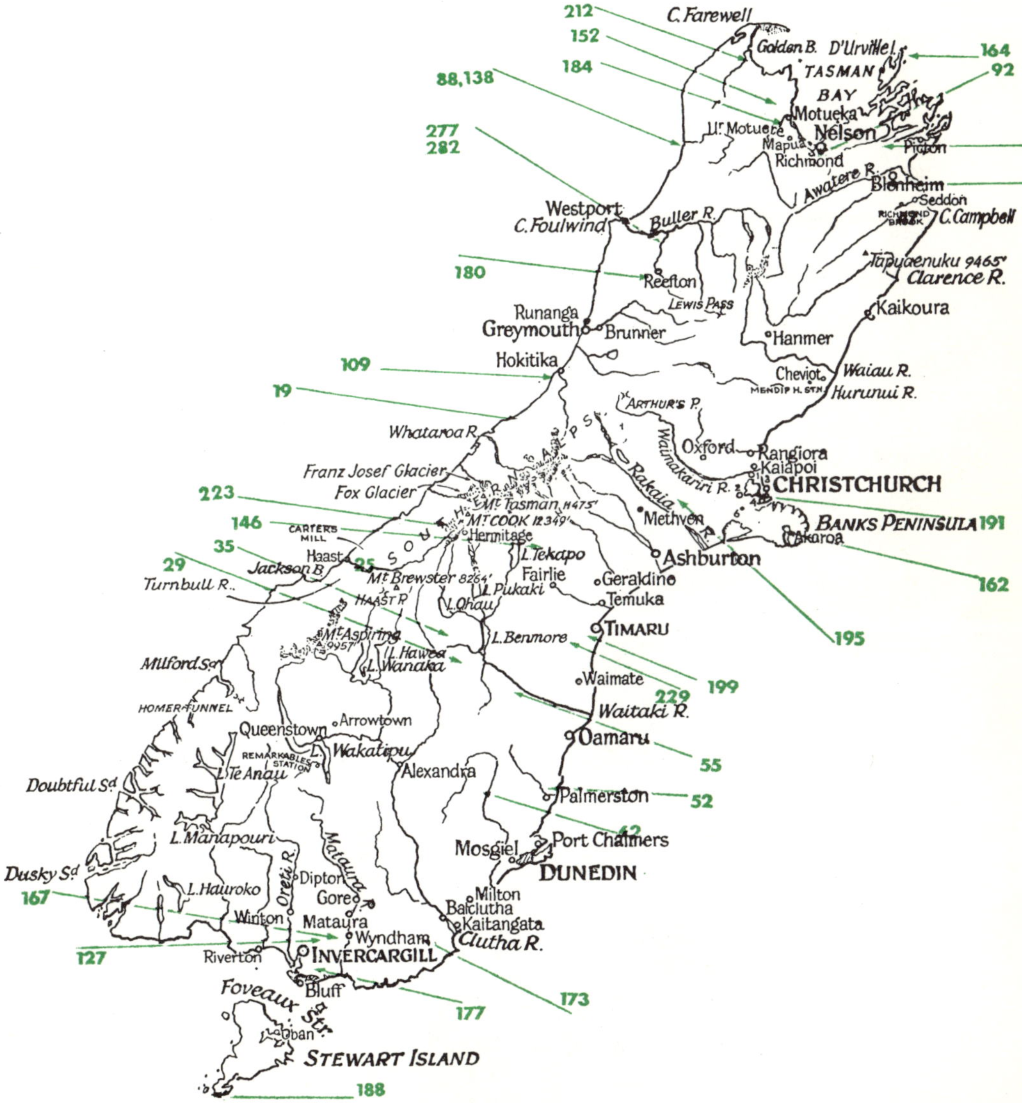

212
152
184
88,138
277
282
164
92
95
142
180
109
19
223
146
35
29
25
191
162
195
199
229
55
52
42
167
127
173
177
188

Open Country Calling

Other publications by Jim Henderson

22 Battalion (NZ Official War History)

Gunner Inglorious (a book of World War II)

New Zealand's South Island in Colour (text)

One Foot at the Pole

Open Country

RMT (NZ Official War History)

Return to Open Country

Te Kao 75 (a district history)

Tobacco Farm (for the Schools Publications Branch of the NZ Department of Education)

Unofficial History (Yarns of Old Soldiers)

Our Open Country

Open Country Muster

To Charlie
With Best Wishes
Betty Kitch

OPEN COUNTRY CALLING

Edited by
Jim Henderson

A.H. & A.W. REED
Wellington Sydney London

First published 1969
Reprinted 1970
Reprinted 1974

A. H. & A. W. REED LTD
182 Wakefield Street, Wellington
51 Whiting Street, Artarmon, NSW 2064
11 Southampton Row, London WC1B 5HA
also
29 Dacre Street, Auckland
165 Cashel Street, Christchurch

ISBN 0 589 00280 5

Printed by Wright and Carman Ltd, Nicolaus Street, Trentham

To the

Pioneers of the Planets

(what Immense Ragwort may Await on Mars!)

and to

the Toilers from the Caves to the Stars

who have Kept the Faith

Greetings

HERE WE COME AGAIN and we all hope you like *Open Country Calling* as much as you did the other two books, *Open Country* and *Return to Open Country*.

Here in this book people who know, writing probably with the main motive of love, tell of our land's roots and the very fibres of our country lives. Hills build character, as well as muscles.

How I wish at times that some of the hasty tourists who scurry through New Zealand could know, could understand, some of this background. I think they'd hesitate a little before passing quick judgments on our national character, our way of life, and our future. Also, how many of us prize "a fair go" beyond wealth and position. This, O overseas visitors, does not mean a dull conformity, or lack of ambition.

"Open Country: People and Places out of Town" first took the air at 1 pm on Sunday 19 February 1961 over the eight YA and YZ stations. The session now plays on sixteen stations each week, and also on shortwave over Radio New Zealand. The number of letters and stories coming in has doubled since television began.

After a year, countryfolk felt they no longer had to apologise for the shameful trade of writing: "As it's a wet day and I've nothing better to do"; "I'm in bed with a bad cold"; "Dipping is over and I've broken my ankle". Later, some admitted it could be as hard as putting up a fenceline: "Say it, yes, but write it? Damned if I can!" And the forlorn admission: "This doesn't seem so good now. It was funny when I heard it, anyhow."

One letter—and the accompanying letters can be as interesting as the scripts themselves—opened: "If I may be so bold as to submit a script which does not concern Karamea or Collingwood . . ."—two brave and rather lonely little settlements especially dear to the session, and often kept in mind when thinking and wondering about listeners. Of course, it's the listener who counts, all the way. Also remembered, as sort of signposts, are special friends and correspondents in Christchurch and Auckland, who miss the countryside, and dream beyond the bitumen.

Another letter started: "It has taken me two years to write this story. I always kept bursting into tears whenever I tried to write it, because I kept remembering the wonderful comrades I had working alongside me." The story appears in this book. Probably you'll pick it.

Nearly all the scripts (fifty-three in one week is the record so far—and only the last three were accepted, with joy, fully making up for the regret in turning down the first fifty) are handwritten, many of them on thin, ruled notepaper, but the words emerge and live, *if* (and this is the key to the whole thing) *the writer really has something to say.*

Scripts which have gone before, in some indirect way, seem to help selection, and set a pattern or standard. The last thing we want to be is hidebound, but there is something to be said for tradition, after all.

"I hope you accept this script. I want to get a telephone installed with the money," wrote one woman, annoying you with her hope. But her script made the grade, then you felt: "Why shouldn't she express her hope? She's being honest about it."

One successful story came in, written on brown wrapping paper!

Another's handwriting was so bad you felt: "Really, this is a bit tough. He should at least have taken a little more trouble." Then came the P.S.: "Please excuse handwriting. I have been all day laying bricks."

Right from the start eight years ago, Auckland and Canterbury Provinces have deadheated for the number of contributions sent in. Southland and Otago remained more or less mute for a good five years, perhaps reserving their opinion. Now, happily, they're well under way in the mail.

A total of 605 scripts came in for a competition in August. The record response so far has been to Tony Nolan's "Help! I'm Lost. What Now?" on page 99. Six hundred people wrote in, seeking copies.

Curiously, some listeners still think a bird's whistle opens and closes each broadcast. It's a high country musterer whistling his dog, as dog listeners recognise, so I've been told.

Once, a script came in about cooking a hedgehog in clay, and eating it. "Untrue," you felt, returning it at once, "New Zealanders just don't do things like that." The very next week, a story arrived from a proved reliable contributor on exactly the same subject. Sometimes you get a wave of stories on trees, then four months later it's suddenly all horses, or telephone exchanges, or whitebaiting expeditions. Quite unpredictable.

You never can tell what Buck O'Connor from Timaru, with his favourite expression "Let the hare sit", will put down on your untidy desk next morning—that's the beauty of it all. And you never know what will be on the air a month from now. But it always comes in, as it has from the beginning.

Never once has the NZBS, and later the NZBC, interfered or made any suggestion, in any way, a wonderful aid.

So a muffled, yet glad thank you all, from this gravel pit of words; and may you enjoy health, happiness, and peace in the weeks and the years to come. Kia Ora.

JIM HENDERSON

Wellington, 1969

Acknowledgments

I am grateful to the New Zealand Broadcasting Corporation for their permission to undertake this book, and to those who have agreed to allow their contributions to the *Open Country* radio programme to be presented here.

J.H.

Contents

Illustrations

The following stories have been illustrated by the artists named

Away With The Archdeacon

THE HOUSE OF PEACE

Archdeacon James Young

FROM THE TALES that are told of exploits in the far south of Westland in "the days before the bridges", you might perhaps think that life was all fun and games. But I can assure you that this was not so, for two sources of nightmare were constantly lurking in the background—accidents and sudden illness.

Just off the beach on the northern side of the mouth of one of the smaller rivers a few miles south of Ross there lived for many years a quiet, friendly old German couple. They had come to the Coast in the days when fortunes were being made on the sluicing claims of the black-sand beaches.

Fabulously rich these beaches were, but all comparatively short-lived; too easy to work, compared for instance to quartz reefs, and therefore soon worked out. And when the gold went, generally the miners went too.

But sometimes a man would stay on, take up a bit of rough farming land, build a house and settle down and stay, often for the rest of his life.

And that's what happened to old Hans and his wife Martha.

When I came on to the Coast they were old identities. Although they were in fact only eight or nine miles from the nearest settlement, they were curiously isolated. You couldn't reach them by any road, only along the beach; the main road south passed them by, only four miles inland, but it was four miles of coastal swamp and heavy bush, so that the road-users would never dream there was anyone there. But to any of us who travelled up and down the beaches, the little house, with its garden and fruit trees, was synonymous with a very kindly and hospitable welcome, a cup of tea and a meal.

I suppose I passed by that way as often as anybody, which wasn't very often, and I came to know old Hans and Martha and to like them very well.

They were a deeply religious old couple, in a quiet unassuming way—Lutherans, whose industry, integrity and charity had won them high respect and affection on the Coast.

They lived very quietly and simply, just two honest hard-working elderly people. They had half a dozen cows which were family pets, with endearing German names. Unfailingly they'd come out of the bush, morning and evening, to be milked; and stand in their turn without leg-rope or halter, while old Hans, squatting on his three-legged stool, would sing to them in an ageing but tuneful baritone, simple German ballads from the days and country of his youth, while the warm milk fell with its rhythmic splash into the foam-topped bucket between his knees. Martha would come across from the spotless dairy, thirty or so yards away, bringing the empty milk-bucket she would exchange with Hans when his was full.

It was her part of the work to carry it back and empty it into the wide milk pans, from which, twenty-four hours later, she would skim the thick yellow cream. Later it would go into the hand-driven "Daisy" churn and be worked up into excellent dairy butter. Old Hans would ride each week the eight or nine miles up the beach and turn in to the nearest township where there was always a ready sale for this particular home-made butter; its quality was so unvaryingly tops. They grew vegetables for sale too, and kept three or four hives of bees; even the bees seemed very much the same sort of family pets as the cows.

Hans, as an old miner, had a thoroughly knowledgable eye for the black sand which, if you are wise in its curious and often tantalising ways, may be very well worth while as a source of income on the Coast. How often, as I've been riding along the beach after a storm, have I dismounted and hopefully panned off a sample of black sand in my sou'wester! But it has never proved auriferous . . . . Old Hans was a much better operator than I.

Riding north along the beach one afternoon, at the top of the tide, I came unexpectedly on the old couple sitting completely worn out but happy quite close to their home, beside a wheelbarrow and a heavy plank and a large mound of black sand just above high-tide mark.

I looked in again three or four days later on my way south. They had made a rough sluicebox with "riffles" across the bottom, and when they streamed down their black sand they took £90 worth of gold from it! They were both big, heavy, slow-moving folk, but I'm sure that wheelbarrow must have run a pretty hot axle that afternoon!

It wasn't very long after that, perhaps six months or so, when on my regular rounds I was riding south from Ross again. It was a hot sunny afternoon and I debated with myself which way I'd take: road through the bush, mill tramtrack, the shortest way, or along the beach. The tide was half out and there was a light sea breeze so I took to the beach. Coming round a slight curve in it, I saw something I couldn't

make out through the slight salt haze: a horse standing and gulls circling round and a couple of logs on the sand. Then I cracked on the pace, for one of the logs stood up and waved to me wildly! It was old Martha with the bridle-rein over her arm and, lying on the sand at her feet, old Hans just as he had pitched out of the saddle.

She was quite incoherent with shock and grief.

"Straight from God you have come," she said. "I have prayed and prayed! Oh the gulls, the cruel, cruel gulls!" And anyone who has seen a stranded fish after the gulls have finished with it will know the kind of picture she had in her mind.

Presently she steadied enough to tell me about it. Hans had set off for the settlement about 9 o'clock on his usual weekly visit. About mid-day his horse came home, alone.

"My heart," she said, "then stopped, for I knew that something was much wrong. I caught the horse's rein and we walked back, following his tracks. Perhaps five miles we had gone and then I saw the gulls circling—the cruel, cruel gulls—and I ran and ran. And there he lay and I was in time, for they had not yet torn him. I tried to move him, but he is heavy, and quite cold. So what could I do? I couldn't go and leave him to the gulls—the cruel, cruel gulls. And no one may come this way for a week. So I prayed and prayed. And the gulls would come closer and closer, with their great beaks and their cruel yellow eyes. And I would get up and throw wood to drive them off. But they would come back and back. And I prayed and prayed. And I thought, wickedly, there is no God who hears. And then, round the turn of the beach in the sunlight, came riding you, straight from God."

I didn't argue about that. But we discussed what we could do. We couldn't get old Hans up on to one of the horses; he was impossibly big and heavy. I was very loath to leave her, nearly hysterical with shock as she was. And I didn't think she was in any state to go the several miles for help. I just didn't know what to suggest.

"Pray with me; pray with me," she begged. "He *does* hear."

So I did pray with her, just as simply and directly as I could. And I explained to Him how we couldn't leave old Hans "because of the gulls", which I knew had become a sort of obsession with her. But I stopped abruptly in the middle; far away off in the bush I heard the chug-chug and rattle of the "Bush loci", a primitive locomotive pulling the tram along wooden rails to bring the bushmen from the "cutting face" home at the end of their day's work.

I don't know how far I got with that prayer, for I stopped abruptly and shouted "Loci!" Martha said "God" very reverently as I sprang on my horse and went galloping away through the scrub and swamp to intercept it at the nearest point. And I got there in time too, and nearly scared the wits out of the loci driver by appearing out of the scrub across the wooden rails close in front of him!

About half the men came back with me and the others went on with the loci to get in touch with the doctor and the police and to come back along the beach with help as soon as they could.

We got old Hans up on to one of the horses and I insisted that Martha should ride the other in a sad little procession back along the beach to the little old house where she wanted to be.

I've one silly incongruous memory from our arrival there. We took old Hans in and laid him down on the couch in the living room. He wasn't a very pretty sight, with his face all twisted and covered with sand, just as he'd pitched out of the saddle. I was standing over him trying to clean him up a bit while Martha went for a clean sheet to cover him. But how much of a fool can a man be, and live? A young gangling gawk of a bushman leaned on the side of the doorway looking in on poor old Han's twisted face.

"Ah!" he said, almost with satisfaction. "It's easy to see he died in agony."

I've no very clear recollection of what followed, but they told me afterwards that I went as white as the sheet, and said very quietly and distinctly: "You'll die in agony if I get my hands on you!" And, putting my hand on the table, I vaulted across it towards him as he stood in the doorway.

I had something of a reputation in those days—quite undeserved—of being fairly handy with my fists, and when he saw me coming his eyes stuck out on stalks and he turned and went pelting out the gate and up the beach "as if the devil was after him", as my informant put it. But it was only me, and I chased him for a few yards to the gate and then stood and watched him, his heavy bushman's boots throwing up little spurts of sand until he turned into the bush about three hundred yards away.

We fixed up all the simple formalities. Martha took me out to the well-kept gardens at the back; a flower garden and a vegetable garden side by side, hers and his, and she showed me where she wanted him buried at the edge of hers, so that there'd be room to bury her at the edge of his.

Two women had come out from the settlement to stay with her through the night and I arranged to stay too, preparing to bed down in the hayshed quite close. Martha asked if she could be alone with her Hans. So, when it was late, I sent the two women off to sleep in the bedroom at the back of the house, and I went quietly out leaving her sitting beside Hans, holding his old cold hand between her two.

I came back early in the morning. No one was stirring, so I looked quietly in. Martha was still sitting as I'd left her, but something in her attitude made me go over and touch her. And her two hands were as cold as his they were holding. I suppose it was all the shock and strain. Her faithful old heart had just stopped.

I went through and spoke to the women in the back room, so that they shouldn't get too much of a shock.

"Thank God," breathed one of them. "I've been dreading the thought of whatever she'd do without him."

When I went back into the front room I noticed that there was a large old German Bible lying open on the couch beside her, from which she had evidently been reading. My German is very sketchy, but it was open at the Book of Wisdom in the Apocrypha, at the third chapter, and then I knew at once what she had been reading. I've read it often as the lesson at an Anzac Day Service:

> *But the souls of the righteous are in*
> *the hand of God, and there shall be no*
> *torment touch them. In the sight of*
> *the unwise they seemed to die and their*
> *departure is taken for misery, and their*
> *going from us to be utter destruction:*
> *but they are in peace.*

The sun was shining on the house as I came out, and high above the front door I saw neatly carved what I'd never noticed before: "The House of Peace."

ON BUILDING A FEED

Archdeacon James Young

"BUILDING A FEED." It's a picturesque Westland term which took my fancy; and I've often been delighted to be on the consumer's end of such an enterprise.

"Come along back to the hut and we'll build a feed."

One of the essentials, of course, is the right kind of appetite, and on the occasion I have in mind there was no question about that!

Three of us—Old Jack and his son, Young Johnnie, both deer cullers, and I—had ridden the thirty or forty miles up the Haast riverbed, arriving at the hut, long since vanished, near the junction of the Haast and Landsborough Rivers. I had been invited to join the two cullers for a day's deer-shooting, and had been fitted out for the occasion in the usual generous Westland way.

I was lent a horse for the expedition: a fine upstanding bay, named Prince.

"He's a good traveller," explained Young Johnnie, "but you need to sit fairly close to the saddle, because if the girth pinches him, or anything, he's liable to go crackers without waiting to explain."

So I sat fairly close to the saddle.

I was also lent a rifle—a beauty; a 303, with ammunition to match.

Off we set up the riverbed, with our tucker in a split sack slung across behind Young Johnnie's saddle. We'd gone only about half a mile when Prince gave a disapproving grunt and turned on his fireworks.

I guided Prince up on to a patch of soft sand where he'd get full value in exercise for his display, and also where a forced landing, if I had to make one, wouldn't be too uncomfortable. With a little encouragement Prince soon returned to a better mind, and we travelled along at a good pace up the riverbed.

Darkness came on, a couple of hours before we reached the hut. Luckily the river was fairly low, as we had to ford it back and forth several times. It swung from side to side of the wide gravel riverbed, cutting us off against the ends of the bluffs. But of course my companions were very familiar with it.

There's always something a bit eerie about travelling a riverbed at night. It's hard to distinguish the grey silt-laden water from the shingle, or to estimate the distance of the dark bush-covered terraces as they blend on either side with the steep slopes of the mountains, rising dimly to a glint of snowy peaks. The valley is hemmed in and full of sound: the run of the river over the rapids; the swell and pause in the roar of the waterfalls; the hunting-cry of the morepork, and the frightened twitter of small birds from a thicket; the weird hissing squawk of the blue duck; the massive thunder of some wild torrent rushing out of a side gorge, such as the aptly named Roaring Billy or the more sinister Roaring Swine. And, through it all, the curiously alien clink-clink of horseshoes on the stones. It all seemed to go on for a long time. And we were glad enough when the horses lifted up their heads and their feet and made determinedly for the little clearing on a low terrace, with the hut hidden away among the trees.

It didn't take us long to make camp. Young Johnnie took the luggage and stores into the hut, while Old Jack and I unsaddled the horses and gave their backs a rub down with a sack. Then we turned them loose in the holding-paddock. There was plenty of good rough feed there, and they set busily to work.

Young Johnnie had the fire roaring and the billy on its way to boiling as it hung on its wire hook from the iron bar across the chimney. In no time we were hoeing contentedly into the cold chops and thick slices of the camp-oven bread we'd brought with us.

They were a delightful pair, these companions of mine. They'd both been born away down in the Far South of Westland, and except for an occasional excursion outside had lived there all their lives. Old Jack was, I suppose, in his early fifties—fairly tall and, like most bushmen, broad-shouldered, spare, and tough. He'd a shrewd, kindly weather-beaten face, and a slightly wary expression, as if he was on his guard against the world taking advantage of his inexperience.

Young Johnnie was about twenty-three, a couple of inches taller than his father and rather slighter, with a pleasant, open, mischievous face, full of fun and good humour. They were great mates, bound by strong bonds of affection which didn't prevent their being constantly on the alert to score off each other, good-humouredly. And when they discovered that I was quite prepared to have them score off me, too—if they could—I came in for my share of the fun.

We were planning to be out round the fringes of the river-flats before dawn, so didn't waste much time in getting to sleep. I heard two wekas talking to each other as they explored round the camp; and further away the shrill cry of two kiwis having a wireless conversation. The next I knew, Old Jack was shaking me awake.

We had a good but swift breakfast of hot buttered toast and cold meat, lubricated by mugs of scalding black tea, before we were out into the first of the dim dawn light with its trailing wisps of mist. We crossed the river, riding very quietly, to where the birch bush fringed the stony flats.

Old Jack stopped and spoke in low, clear tones to Young Johnnie—though I knew very well it was meant for me.

"Now remember, Johnnie, never fire unless we can be sure to kill. I hate leaving a wounded beast to suffer," he added, turning to me. "If we wound a deer we follow it till we get it, if it takes us over into Otago."

"I'll be careful," I said, "which means I won't fire many shots."

"Oh, you'll be all right," he said, "if you can get a shot out in the clear."

We had a great morning's shooting. I thought I knew a little about handling a rifle, but I simply wasn't in the same street with these fellows! It's one thing to shoot a tame jam tin off a stump, but quite another to fire from the saddle and shoot a hind in the back of the head when she's galloping away through waist-high scrub!

They didn't by any means fire at everything they saw; as cullers their purpose was to get rid of the malformed and the scrubby, poor specimens. I don't know how many we shot, but by about mid-day we'd been round all the handy flats and all the deer had retreated deep into the bush. So we started to move homewards. We were on the same side of the river as the hut—we'd crossed and recrossed several times—and perhaps a mile away from it, making our way peacefully along.

"I'm getting a bit hungry," said Old Jack.

"Hungry!" exclaimed young Johnnie. "I'm that empty my navel's chafing against my backbone."

"Empty, are you?" grinned Old Jack. "Come on, then—race you back to the hut and we'll build a feed that'll put some stuffing into you."

As Young Johnnie got away to a flying start, Old Jack was not more than a couple of seconds behind him, and away went the two lunatics cutting each other out and running each other up against piles of driftwood and such delights, galloping madly over the big round stones of the riverbed where none but a river-bred Westland horse could make such pace without breaking his own or his rider's neck!

Back at the hut, we set to work—after a hunk of bread and cheese (just to prevent further chafing)—to "build a feed". We lit a proper camp-oven fire to roast three paradise ducks, about the size of small geese, together with potatoes, carrots, onions and all the doings. I earned great credit by finding on the riverbed a fairly big and bone-dry lancewood tree, the best of all fuel for the camp oven. It quickly burns down to clear, smokeless, glowing embers, which keep on glowing until they fall quietly into pure white ash. You stand or hang your camp oven over the embers, with your ducks etc. inside, plus fat and water, and heap your embers also on the lid, which has a turned-up edge. And oh, the fragrance which begins to ascend like incense!

And at last the feed was built, and we sat down to a large plate each, with our roast duck piled high with trimmings. Then followed comparative silence while we dealt with the business in hand. Old Jack was a notable trencherman, and when Young Johnnie and I had already reached saturation point Old Jack was still going strong. We kept plying him with further delicacies, and looked on with a kind of awe.

At last he pushed back his plate. "No more, thank yer, Johnnie. I guess I'm about blocked."

"I was just thinking you must be pretty near blocked," mused Young Johnnie.

"I think I'll just top off with a bit of spotted dog," said Old Jack, and, wiping his skinning knife on the seat of his pants to sterilise it, he proceeded to cut off a great slice of the solid, doughy, currant-splattered spotted dog—eight inches by two—and wade steadily through it.

Young Johnnie, in a stage whisper, confided to me: "By cripes, the old man's not as near blocked as I thought he was!"

But it was a noble feed that we'd built.

A NIGHT IN THE TUSSOCK

Archdeacon James Young

HOW STILL it is here in the tussock.

All day, the whirr and scrape of the wheels has been in my ears, and the rattle and bounce of a stone flying off from the tyres as the trusty old bike covered the miles on these loose-metalled roads. The going has been easy since turning, after mid-day at Omarama, out of the Mackenzie Country's dust into the richer tussock downs of this Waitaki Valley. For the wind has been behind us, the old bike and I, and the trend of the country is downhill. And now the sun is setting behind the great peaks.

How I wish I could paint, but how this would break the heart of the artist. It's never the same for three seconds: the blaze of the last of the sun, then the fire that's gilding the peaks; and the gorges are filling with shadows, pink tinges then to purples and blacks with a flash of the snow in a rift. And the sky, backing the peaks, from golden to rose and at last a green so fragile, transparent, unearthly that it makes to perfection the background for the jagged, menacing peaks with their dog-toothed implacable line . . . And here come the stars, through the fast-gathering darkness.

I'll be glad of this shelter I found. Just a dip, like the cup of your hand, and the tall tussock round, and a place to lie out of the draught of the wind going over. And it's no trouble to cut with a sheath-knife the dry, tawny tussock to make the best of all beds. I'll be glad of it too, for the wind will be gone in an hour and the frost will be hoary and stiff up here, on this mid-autumn night. So I sit with my swag at my back, and puff a contented last pipe, and listen—to the stillness.

A little wind is just stirring the tussock, with a sound like nothing else I know, a small gently-rasping hiss; not a menacing sound, but warm and contented and friendly. It sounds as it smells, sun-warmed with a tang of the heights and the wilds. And it smells as it looks: tawny, untamed, with the waves of the wind running over it, on the flanks of the mountains and the downward flung roll of the downs. Yes, the tussock; it's faithful, consistent. Everyway that you meet it, in sight and scent and sound, lovely without being striking, shy without being wild, friendly without being tame.

Through the stillness, a morepork is calling, drawing closer and closer, not disturbing the silence with his half-plaintive, half-mournful lost call, as of one seeking friends. Suddenly he breaks into his strident hunting cry. I hate it, and you can almost feel the small creatures cower and hide at the base of the tussocks, the fieldmice and larks and young rabbits. He's flying right over me now, and I sit up and suggest he departs, which he does, a dark shadow on startled ghost-whispering wings.

A buckrabbit at the mouth of his burrow thumps warningly with his hind feet—a wireless message to all who can hear it, of danger about. I wonder which he thinks dangerous, the morepork or me?

Far off a pukeko calls, an eerie sound unless you know the cheery, cheeky customer, purple-coated, red-beaked and long red legged, with that jaunty ridiculous flick-flickering little white tail. There must be a swamp over there, at the foot of those low-lying hills.

But the night's growing chill. I unbend my swag and wrap myself up in my blankets, and wind over all the old tent-fly to make a cocoon against draughts and the damp of the dew and frost, with a flap of it over my head to keep the cold off—off, well, the place where the curls used to be. I take a last peep at the stars, very silent, serene, and wish my ears could be tuned to the song of the spheres.

Yes, it's cold; and I'm glad I'm old camper enough to have chosen my site facing east, so that the sun, when he tops the first roll of the downs, will shine full on my cocoon.

There's that cheeky pukeko again . . . and I bury my nose in the tussock—sun dried, sweet-scented and tawny. . . .

And before you turn round, here's the morning.

Half-an-eye, from beneath my head-cover, sees the sun coming over the rise. The frost sparkles like jewels on the long drooping leaves of the tussocks. I'll doze a bit longer, snug in my cocoon until the chill day's a bit aired.

What a funny small sound: a small hopping and scratching. From where? Why, it sounds up above me. With the slightest of movements, I pull back an inch of the covering tent-fly—and there, sitting up half way down my cocoon, is a little grey rabbit, with his breast to the sun, enjoying the warmth and preening his whiskers with little fore-paws! What a picture, and the sun making almost a halo, shining through his soft fur. I lie like the log that he obviously takes me to be—and not a bad guess, I'd been sleeping like one.

But soon, with that sixth sense of the wild, he's uneasy. His hair-

dressing stops. His ears swivel round, and he semaphores signs of alarm.

It's a shame to disappoint him, so with a wild "woof" like the biggest bad wolf, I sit up with a jerk. I've a split-second's glimpse of a pair of protruding brown eyes, like two half glassy marbles, and then, with a spring that must surely have broken the world-record sitting-high-jump, he's away and I see his white scut flicker into the mouth of a burrow on the upward roll of the down.

Monotonous country, these rolling brown hills? Not a bit of it—with no two tussocks alike. And full of surprises, as that little rabbit can testify. What a tale he'll have to tell his grandchildren of the morning he broke the world record for the sitting highjump; and won't they just think him the Prince of Liars, a calumny all too commonly the fate of us story tellers!

The Heading Dog Who Split in Half — and Other Sheepdog Yarns

THE HEADING DOG WHO SPLIT IN HALF —AND OTHER SHEEPDOG YARNS

Bill Timmings tapes Mackenzie Musterers

THIS DOG STORY was told to me almost thirty years ago, when I started mustering, by a fellow who'd had it told to him many, many years before. The story concerned a dog: a very fast heading dog that had become a legend by the time he was three years old. This dog was reputed to be so fast that when he was running out to head a sheep he could not be picked up with the naked eye.

A certain boss in the Mackenzie Country took a gang of musterers out to muster a very wild piece of range. This country was shingly and it was high. To get the big strapping Merino wethers off this rough country they had to get out on the evening before, so the sheep would be unaware of their presence, and strike very early the next morning.

So the boss led his men out and they camped in the tin huts below this range. Unfortunately for the muster two or three young dogs in the various teams barked at night. The old Merino wethers heard them and got restless. They mobbed up during the night, and when dawn came the musterers saw Merino wethers stringing out through the shingle saddle onto what was known as no-man's-land: country that didn't belong to anyone.

The boss immediately broke down and wept; with these Merinos stringing away through the saddle there was only one man who could save the station: the shepherd with his famous heading dog. He set the dog a line across the flat before it reached the range, but very unfortunately for the dog someone had driven in a standard—a fencing standard which was concealed by tussock and snow grass—so this very fast heading dog, with terrific acceleration which threw the tussocks and the rocks behind him as he got motoring, struck the standard cleanly on the nose, and it split him completely from his nose to his tail. The owner of the dog, seeing his dog lying in two perfect halves, rushed up, grabbed the two halves, and lo and behold! within a matter of moments, the dog was on his way again.

Several hours later the dog appeared with the sheep; but now the musterers saw that when the dog's owner, in his great distress, had picked up the two halves, he'd slapped them together one half the wrong way up; and there was the dog running on two legs—and two legs sticking up in the air.

But wait; this isn't the end of the story. It didn't spoil the dog—no—it *improved* him: when he got footsore on two feet, the owner simply turned him up and ran him on the other two feet. So he always had a fresh dog.

I've a dog story to tell you of a well-known dog-trialist who had two nice pups to give away to get broken in. He gave one to the parson and one to the village idiot. Time went on and the village idiot trained his dog to be a beautiful dog: it would work and do almost anything. But the parson just couldn't get anything out of his dog at all.

So one day when he met this nit in the village, he asked: "How is it that you can get so much out of your dog, and I can't get mine to do anything at all?"

"Oh, that's easy enough," said the village nit. "You've got to have more brains than the dog."

Once we were mustering Cass Hill and Sugar Loaf on Grasmere, and I was on the top. The Cass Hill is a very round bulging hill covered with tawine scrub. One of the musterers below me called Luke Buchanan was coming along and he saw nineteen sheep going back behind him. So I sent my old dog after them. When I caught this chap up, I asked him to go back two spurs and see if he could see the dog but there was no sign of him. We had to go on without him.

We went right on to the saddle between the Cass Hill and the Sugar Loaf (there's a pretty steep pull straight up to the top of the Sugar Loaf), but the chaps of the Waimak side were held up. We had to wait there for about an hour. We walked as far as we dare up the Sugar Loaf and sat down, and in about an hour's time, my old dog got the sheep into the saddle and he sat there; and we were sitting down quietly with the dogs asleep, and he sat there and then he came up through the saddle with the sheep: all nineteen of them. He's wandering along behind them looking at the sky like he used to, and he turned the sheep round on to the right side of the hill and I called him up to me.

This chap Buchanan had mustered with me for several years. He called my dog up to him; he gave him half his lunch and put his hands round his neck and cried.

After the First World War a soldier-settlement grew up on the other side of Lake Pukaki. Money was very scarce, and horses were the settlers' only means of transport. Every Saturday afternoon the chaps used to ride down to the village to get their stores and have a gossip, and naturally they'd finish up at the old Pukaki Hotel, their dogs trailing behind them. One week Bill didn't turn up; everybody was concerned until somebody said they'd seen Bill on Saturday morning so he must be all right. The next week there was still no sign of Bill, but again somebody had seen him that very morning. "He seemed quite all right; I'll take his stores home."

And another week went by and all the other settlers turned up as usual, but still no sign of Bill. Eventually somebody walked out the front door and had a look up the road and saw him coming. So they gave him another five minutes and while Bill tied the horse up everybody crowded out of the hotel, shook hands with him and generally made quite a fuss of him.

Eventually somebody asked: "Why haven't you been down, Bill? What's been wrong with you?"

Old Bill said: "My dog's been sick."

"But surely he wasn't so sick that you couldn't come down to get your stores?"

"Oh, I suppose I could have come down—but a man would look a fool coming down without a dog."

Darkie was a black heading dog I had for six years, and a very good dog he was. We were mustering on Haldon: on the Kirklesson block where they run their wethers. I had the beat Scrubby Gully, where there was a lot of woolly sheep; I hunted Darkie on a mob and away he went. But he didn't come back with a single sheep and I had to go on without him. When we got to camp that night Darkie was missing and I was very worried about him. The weeks went by and still Darkie didn't show up.

Two months later we went back to muster the same block, so I asked Ian Innes the boss to give me the same beat I had on the last muster.

"I may see some sign of Darkie, even his remains, that would satisfy me."

When I came to Scrubby Gully I had a look round for Darkie, but there was no sign of any dog. So I called him, and out came the noble Darkie from the creek: his coat was shiny and he was as fat as a seal.

I thought, "You old so-and-so, you've been killing your own mutton: I'll have to get rid of you. Darkie, I never though you'd do a thing like that."

Anyway, away we went and I came to a fence; I thought I could see something on it. When I looked again I saw it was sheepskins, so I said to Darkie, "You're not so bad after all, Darkie: you've hung the skins on the fence."

There was half a dozen of us sitting around the breakfast table: three o'clock in the morning and a day's mustering ahead. The boss was asking us what sort of dogs we had and if they were good dogs. Yes, we all had good dogs. So away we went after breakfast, to walk the eight miles or so up the hill. About seven miles up, the boss turned around.

"Good heavens," he said, "where are my dogs?"

They were still on the chains at the dog kennels, and back the boss had to go!

Of all the dogs I've owned I best remember Scott.

I was fifteen and just finished ten hours harvesting for a farmer at half a crown an hour. However, it suited me fine when he asked me to accept a dog in payment, because I was about to go to my first shepherding job. So I took this dog—a "pig in a poke" more or less, although of course the farmer reckoned it was from a great breed—and away I went.

Even while the head shepherd was showing me round the new place Scott rounded up the first mob of sheep in the paddock where we were riding, and as soon as he'd got them he careered into the next paddock, and so on. I had a very busy morning collecting my dog. Within a fortnight, with constant work, he was a great leading dog, a huntaway and heading dog combined.

The war broke out shortly afterwards and I went into camp. I was so fond of this dog I decided to leave it with my family in Geraldine until I came back. I've never forgotten the weekends when I came home from leave. We would come from Burnham to Orari in the train, then to Geraldine in the bus, and the last five miles in the family gig. In those days not every farmer ran a car—we were just coming out of the Depression. Waiting for the family I would always see first this little black spot on the road, a mile or so out in front of the gig. Scott knew I was coming, and great would be the reunion when he met me.

Shortly afterwards I went overseas, and this dog lay around at home, not following anybody, and refusing to work for anybody else. Many weeks after he picked on one of my brothers and stuck to him until I came back two years later. I couldn't say the dog was expecting me back, but I still remember the reception when I got into our back yard at home. I don't think I got a warmer welcome from anybody.

Many years ago I was head shepherd on Grasmere Station. I was sent one morning, with a young fellow who'd been on the place only about five days, to catch sheep round the bottom of a block called Bailey Block, along Lake Pearson. This country had only just been taken over by the station off Craigieburn and there was a lot of rough sheep on it. It's a big hill, too, running up to five and a half to six thousand feet and rough and shingly at the top. We were just ketching round the bottom.

It was blowing a very heavy nor'-west wind and we stopped in the shelter of a little knob to have a smoke. Looking up we could see about twenty-three double fleece sheep pulling out through a saddle right up in the shingle.

This young fellow said to me, "It would be a good dog that could catch those sheep."

"Well," I said, "I've seen dogs that could do it."

"But you've never owned one."

"Yes, I have."

"Well," he said, "you haven't got one now."

This got under my skin, so I sent a dog that I had—a dog called Bounce, a pure white beardie dog—and he went away up and caught the sheep. He didn't come back with them, so we left him. He was a good four miles from the homestead and there was the Ribbontree River to go through.

We went on and finished off round the station. Andy Taylor was manager of the station at that time and my wife was cooking there. We had a cup of tea and were talking about the station work, so I said to Mr Taylor: "I've done a foolish thing this morning. I sent Bounce after some of those rough sheep in the paddie and he hasn't come back."

At four o'clock in the afternoon he arrived at the station with the sheep, and the manager of the place went out across the downs and took the sheep off the dog and gave him the shoulder of meat we were going to have for tea.

The year before I was at Double Hill there was a musterer there called Kit Anderson. Now on Double Hill out at Cummings block there's a place they call the King Drive: it's very bluffy, a very bad place, with only a tiny sheeptrack through it. Almost everyone going through there gets bluffed, but the other musterers didn't use to worry about them until they had mustered off, when they'd go back and help them out.

Well this boy Anderson must have got hit on the head with a small stone and knocked down and become sun-struck, because when they went back to find him, he wasn't there. They searched for three days and three nights. Finally Mr Pollock found him on Mt Hutt station. He was still travelling on his hands and knees, and his tongue was out, black, as big as his mouth. Some of his dogs had come back to the camp, but two of them were still with him. They had a terrible job to catch these two; and the dogs wouldn't allow them to go near the man.

Two of his dogs remained with him in that country where there was no water, for three days and three nights, and were still with him, trying to protect him: wouldn't allow anybody to go near him.

AND UP NORTH

John McCaw

THE EAST COAST is the place for dog stories. There never were such clever, versatile, resolute, gallant, intrepid, valiant, sagacious, and altogether remarkable dogs as those bred on the big stations on the Coast.

My experience of them did not bear out these tremendous claims. I saw a few good ones, certainly, but the main bunch were pretty awful brutes. However, none could compare with Keith Wyllie's Mac we saw every week at Frankton.

Keith was a dealer in stock of all types of cattle and sheep, and he lived a couple of miles away from the saleyards, over the main railway crossing and just beyond the Tuhikaramea crossroads. Keith would sort up his sale stock and send Mac off with the first mob while he got a second lot ready.

Mac would guide his mob—he didn't mind whether he had cattle or sheep—past the busy crossroads up to the equally busy railway crossing with its three sets of tracks. Here Mac would scout ahead, make sure the crossing was clear of trains, hurry his charges over and around two corners to the saleyards: cattle to the cattle entrance which was down an alley; or, if he had sheep, to the sheep entrance. If a second mob were on the road, the dog would keep the two lots from boxing. When the gate was opened for him by one of us, Mac would return to his boss for more instructions.

Then Jimmy Ashdown at Te Horo had a wonderful little dog with whom Jimmy, a bachelor, held long conversations—a bit one-sided, but the dog appeared to comprehend all Jimmy said.

I arrived a bit ahead of time one day to find Jimmy having a siesta and no sheep in the yards. While pulling on his boots, Jimmy called his dog over to his knee and told him quietly that I had come to buy his wethers, and it would be a good idea if he, the dog, went and mustered them up to the gate.

Away went the dog.

Jimmy and I strolled over a couple of little sandhills to find the dog holding a small mob of sheep at a gate in the corner of a paddock.

Jimmy gave a wee whistle, the dog came at once, whereupon Jimmy pulled his ear playfully and said: "You stupid ass, I said *wethers*—not ewes."

The dog gave a grin—dinkum, a real grin—then dashed off into another paddock and, without further instructions, collected up the wethers.

A different sort of dog was the foxie that lived right opposite Feilding saleyards with his owner, an elderly widow with an acid tongue. The widow and the dog lived together in a little cottage only recently demolished to make room for the county offices. This fox terrier was a terrific nuisance to the drovers bringing sheep into the front entrance to the yards. He would sally forth and harry the sheep, or stand in the gateway of the yards just when the head of the mob was turning in.

Very early one summer morning Norman Gorton, a leading auctioneer in the Manawatu, was helping a client drive his big draft of ewes through the township and into the yards. The ewes from high country were as timid as sambur deer and nearly as fleet. The mob had baulked at the saleyard gateway—it was new to them, of course—but were just starting in, in nervous fashion, when out bounced the foxie and headed them off.

As the ewes fled in three directions, Norman shouted the foxie's pedigree and looked round for a weapon. A horseshoe was the first thing to hand and he let fly with it, caught the foxie on the head, and killed him stone dead.

Norman had had many a run in with the old girl who owned the dog, and as he hadn't always had the best of it he wasn't looking for one on this account. He picked up the corpse, thrust it under his overcoat, and cast about for a more permanent hiding place.

He noticed a great cloud of smoke rising from the nearby bakehouse, so he slipped over to the kitchen, sent the baker into the front shop to butter half a dozen buns, and then thrust the late lamented little dog into the blazing furnace.

This is the first recorded cremation in Feilding.

The Denbigh Hotel lies over the road from the bakery, its proximity to the saleyards making a lucrative stand but also one fraught with some danger.

Roy Shannon, when driving his sheep into the yards, broke the local bylaws by coming across the Square and past the hotel—a forbidden route. It was only just daylight, with no one about except the hotel porter who, having scrubbed the main entrance and passage, was leaning against the door post having a spell.

An athletic young wether broke away from the mob and dived down the alleyway leading to the rear of the pub. Roy Shannon sent off a couple of dogs to bring the sheep back. Two more dogs from the saleyards joined in the chase but, having had no orders, weren't sure whether the wether was coming or going.

In the turmoil the poor bewildered sheep bolted into the hotel, down the passage and into a store room half-full of empties—and soon the other half was full of dogs.

Shannon's dogs objected to the strangers taking part in the hunt, told them so in howling rage and, forgetting about the wether, sailed into the strange dogs, banging the door shut in the scrimmage.

Pandemonium is only a mild word for the row that followed.

What with the barking and howling of dogs, the shrill whistling of their owners, the falsetto shouting of the Italian porter, the crashing of crates, the smashing of glass, the language of the suddenly awakened proprietor, and the slamming of doors as boarders rushed terror-stricken away from the earthquake—well, Feilding hasn't ever had such a stir since.

A Great Game, This Farming

SNOW IN THE HIGH COUNTRY

David McLeod

AS A SHEEPFARMER I'm in duty bound to hate and fear snow and yet in all this world there can be no landscape more dazzlingly beautiful than the New Zealand high country after a good fall. Where we live the mountain beech—which all high-country men call "white birch" —clothes many of the spurs, moulded to their shapes like a dark green cloak flung carelessly across the shoulder of a girl. After a snowfall the branches are etched in silver and the low sun strikes slanting across, too winter-weak to thaw the rime.

All this is fine when the fall is moderate—perhaps six inches on the flats and a foot or two at 4,000 feet. If it's not too early in the winter the direct rays of the sun on the steeper slopes will soon begin to open dark patches where the tussocks break through. Good healthy tussock is a great protection in snow, leaving hummocks and hollows and preventing the formation of a solid blanket which is so difficult to thaw. Early in the winter when the sun's at its lowest, the flats don't clear readily because of its oblique rays, and often in June you have the rather strange contrast of hillsides largely open while the flats are smooth and white.

This is when true "winter country" can be seen, and without a good proportion of this winter country, no high country run can be regarded as "safe". Aspect is everything—aspect and steepness—and this is where the conflict comes in between the sheepfarmer and the soil conservator. The very country so vital for safety in winter is that most vulnerable to erosion.

When the snow is really deep, say a foot or more on the flats and belly-deep up on the hills, we can't let Nature take its course. Something's got to be done, and quickly too.

In the old days there was nothing you could do except ride or walk to where you thought the sheep might be and, when you got there, tramp tracks and dog the stranded sheep out to easier country. If that country didn't clear, the sheep were still little better off, because few people had anything to feed them on. That was when the really big losses occurred.

Now compare what happened in June and July last year. Helicopters were used in the Queenstown-Wanaka area to take both men and hay to the stock, and there was hay available to everyone. Certainly the stock had a bad time: there are stories of cattle losing their hoofs from tramping round in the heavily crusted snow, and horses with raw and bleeding fetlocks. But in spite of all this there doesn't seem to have been serious losses, and though wool will suffer, and perhaps the lambing percentage, no permanent harm will have been done.

It's true that times have changed and there's no longer much danger of very large losses today, but we've learnt a few lessons from this winter none the less; or at least we ought to have. First and foremost comes the rule: *there is no substitute for hay.* Silage can't be transported any distance, and crops are mostly buried. Those who can grow turnips are better off than those who rely on cereals or saved grass; and those whose land is good enough to put in a bit of chou moellier with their turnips are better off still. I'll long remember the pleasure I got from seeing 2,000 hoggets feeding happily on the stalwart stems sticking up through a foot or more of snow.

All these crops of course depend upon what land you can cultivate in the high country, and the farming organisation to support it. Many properties have little suitable land and many too low a rainfall. Most high-country places can find enough land to grow some hay, either lucerne or clover, and if they can't, it can be bought and transported. Hay can be fed on the ground to stock whether they are snowed in or not, and if you're desperate it can be carted out by plane or helicopter. I heard one amusing story a few years ago of a very energetic runholder flying out hay for several weeks to sheep snowed in on his far boundary. He dropped the bales from the aeroplane and watched with great glee as the hungry wethers gathered round to devour it. He was somewhat less delighted when the sheep were finally rescued, to find that they all belonged to his neighbour!

Another thing we should have learnt from the trials of this last winter is the immense *variety* of snow. It's not just a lot of cold white stuff all with the same texture and properties. Some snow is wet, some is dry, some very cold, some even relatively warm. Sometimes it forms a crust and sometimes it doesn't. Sometimes it binds and sometimes it slides. How can we recognise these variations?

In our district the first fall last year was very wet: from the twelve inches which settled on the ground we got four inches of water in the rain-gauge. (This was the fall which sent an avalanche thundering down onto a new ski lodge.) It lay solid and heavy on the hills and flats, granular in form like snow in spring. The first snowfall was followed by severe frosts, below zero in many places and minus 14° in another spot. This sharp cold formed the wicked crust—the freezing of wet granular snow. Then came another fall, quite different: the dry crystalline snow so typical of mid-winter. A foot of it thawed down to only one inch of water, and it sat like an insulating blanket on top of the crust below. Never have I seen so many avalanches in winter, some large, many quite small; everywhere you looked, the soft slippery stuff was sliding off the hard surface beneath.

Ski-fields were closed and shepherds looked anxiously at the slopes above whenever they had to go on steep country. The losses from sheep being carried away and smothered in these slides, only became known for sure with the shearing musters later on.

I've never seen so much damage in the bush. Avalanches have torn into it, trees fallen higgledy-piggledy and, most curious of all, in some places near the upper bush line it looks as if the weight of snow had stripped the branches off, leaving the trunks like bare poles. Several places have reported keas at work on the trapped sheep and the old argument springs up: what makes keas suddenly start to kill? Is it hunger? Is it mischief? Is it the golden opportunity presented by the helpless animals? To us the reason doesn't much matter, it's just another way of losing more sheep.

All this brings home to us the importance of understanding snow. For many years the winters have been relatively kind, and almost a generation has grown up which has never had this battle to fight.

But in 1939 two of us spent a whole day walking on the solid crust on a big north-facing hill. The snow was feet deep under our boots and huddled mobs of wethers sheltered in little clumps of bush or on rocky spurs. There had been sun to thaw the surface and frost to harden it, and we seized the opportunity of a sunless day to force the sheep out of their refuges and drive them in long strings down the hill to safety. Once they found the crust would bear them they were only too thankful to go. That day we saved hundreds of sheep, but one break in the clouds or a change of wind and we and our wethers would have been waist-deep.

In the same winter I missed the picture of a lifetime. I used to carry a movie camera tucked inside my shirt, so that it was always handy. We had found about 300 wethers in a patch of bush where they had gnawed the bark as high as they could reach after eating every living thing at ground level. To reach the lower country we had to cross a large lagoon—frozen, of course, and covered by about three feet of snow. For hours the two of us tramped to and fro like mediaeval criminals in a treadmill, packing the deep loose stuff until there was a track firm enough and wide enough to take the sheep in single file. The track we had to tramp was about 400 yards long, and I can tell you our legs were weary—we couldn't use our horses for fear of them breaking through the ice.

With great difficulty we found a leader bold enough to set forth alone across this strange pathway. There's usually a strong intelligent wether in a mob if only you can find him. Anyway, with much persuasion and dogging on each side we got our 300 sheep out across the dead white flat. They were practically head and tail across the whole distance with only their woolly backs just showing above the snow. It was dazzling in the sunshine and the sky above was blue and cloudless. On the far side lay a fringe of dark green bush and the hungry sheep saw their salvation ahead and plunged forward. I reached for my camera—and it wasn't there. I'd left it for once at home.

That camera was the cause of another kind of mishap. I had gone out alone on skis to look for sheep on the Burnt Face, our old wether block soon to be retired from grazing by the Forest Service. I found a hundred or two and was able to drive them down from the far boundary of the block, getting some splendid shots of sheep and dogs as I came. The snow was so cold that day that it froze in huge snowballs on my long-haired dogs, and I had to cut the snowballs off with a knife when the dogs could no longer drag themselves along. Halfway home I noticed that my shirt was hanging out and my camera was gone. Putting sealskins back on my skis to stop them slipping as I climbed, I set off back along the track, well marked by the trampling feet of the sheep, and late in the afternoon I found my camera. When I finally reached the flat I had a ten mile ride to get home and night had fallen.

By this time my anxious wife had not unnaturally come to the conclusion that I was lying injured in the snow in some inaccessible place, and was by now quietly freezing to death. She had gathered all the men and organised a search party when on the clear frosty air they heard the unmistakable clatter of my horse's hooves on the concrete bridge nearly three miles from home.

I had enjoyed my day, skiing alone in the sunshine above the mundane world, successfully doing a job that I loved. In the selfish confidence of youth I had hardly spared a thought for those who sat at home wondering where I was.

THE FLOODS CAME

Marion Trotter

FOR FOUR YEARS there had been a drought in North Otago. Four seasons with very little rainfall. Four seasons watching the failing crops, the parched countryside and the poor, thin stock.

Some farmers invested in expensive irrigation plants, others sent their stock away. Almost every day, large trucks of sheep would pass our gate on their way south, and large loaded trucks of hay would be coming up the road going north.

The inland farmers were the hardest hit. Most of them sent their stock to South Otago's luscious pastures, and the owners nearly turned green with envy when they saw the ewes grazing, their legs hidden by grass and clover, like discreet Victorian ladies.

For us, near the coast, it wasn't quite so bad. We'd had a fairly reasonable spring. Often it rained a little at night, but in the morning the persistent wind seemed to dry up every drop of moisture. However, the crops held their own, or almost, and the stock wasn't so bad either.

But early in the new year it was hot and dry. The paddocks were parched and brown, and when the wheat turned colour there wasn't a green thing to be seen. The only topic of conversation was the weather. All eyes turned to the cloudless sky, and that evil little north-east wind was with us almost every day. The creek, which runs through the farm quite near the house, was very shallow all summer. There hadn't been a flood for so long that decaying vegetation and slimy weed covered all the good swimming pools, and the boat was left tied up most of the summer holidays.

I got so used to the idea that the creek would keep within its banks for evermore that I started a new garden in the gully. John brought in a load of lovely mountain rocks, and I got started. I planned wonderful things: crazy paved paths and a rocky pool. We planted ornamental flax and cabbage-trees on the slope, rhododendrons at the bottom, and made a sunken iris garden. It would be years before it really looked like the picture of my dreams, but it was a good start.

Towards the end of February the wheat was ready and harvesting began. It was lovely not having to worry about the moisture content of the grain, and the men worked early and late. But only half of it was in the bags when the rain came. Thursday dawned damp and misty but on Friday, without any warning from the weather office, it rained all day. Softly, quietly, it fell on the thirsty parched earth.

That morning, as I walked around my garden, I noticed how much fresher and greener everything looked. The lilies, the pride of the garden, were just opening up. I'd spent many evenings watering them with the hose, but this was the real thing. The pink waxen petals unfurled as the gentle rain drops fell on them. They were more than beautiful—I just can't think of a word to describe their beauty.

On that Friday we went to town, as I think all North Otago country folk did. It was extremely hard to find parking space, and I've never seen so many smiling faces.

During that night it rained really hard and rained all night; we smiled in the dark, happy with the knowledge that our drought was over.

By daylight all we could see was a raging river. We were completely surrounded.

From the back door we saw that the bridge had already gone. Great trees, the horses' water-trough and a chicken coop went sailing by. It was really terrifying. Then a red and yellow object floated past—the mat from the boys' hut. They were asleep, then had to wade across to the house through two feet of muddy water—which was still rising.

I rushed to look out the front window. The shrubs were already half covered, the flower beds were about three feet under. I could just see the little red flag on top of the mailbox. Pink spots surged before my eyes. I thought I must have become dizzy and was seeing things—but no, they were there all right, some bobbing about in the dirty water, some caught up in the tree-tops. Then I realised that the pink spots were the lovely lilies I'd been nursing so fondly all summer.

Our chief concern was the sheep on the flats. We didn't know if they would move to a higher level or not. The dogs were all tied up to their kennels on the other side of the raging torrent, and we couldn't do a thing about it. Luckily a neighbour got to the dogs in time. When the water started to lap over the doorstep, we lifted the lighter furniture on to beds and tables, pencil-marked the water level on the back door, made a nice cup of tea, and waited.

We didn't know how much higher the floodwaters would rise. The main road was completely blocked, and neighbours were arranging for tractors and a boat to get us out. However, next time I looked, the water was below my pencil mark. Thank heaven for that. . . .

We weren't housebound for long, but it was days before we knew the extent of the damage to the farm.

There had been five and a half inches of rain in ten hours, and probably twice that amount in the hills. The slow little creek had turned into a dirty, swift-moving river over night, sweeping all before it towards the sea. Fences were flat, great trees uprooted, tons of good soil washed off cultivated paddocks, and bridges gone.

One shed had disappeared altogether, leaving only the flat concrete floor. In another place shingle and stones covered half a grass paddock, and carcases of sheep hung up in the willow trees. A walnut tree, planted nearly fifty years before and bearing its first really good crop of nuts, was uprooted, and of course our boat was gone.

Because of the low wool price, many farmers here have turned to crops of wheat or barley, in spite of the risk of overproduction and a fall in price. Most of these crops had not been harvested when the rain came and they sprouted, in some cases becoming a total loss.

Now the countryside is clothed in a vivid green. There is ten times more feed than the stock can possibly eat. The trucks have started running again, bringing the sheep back from the south.

My poor sunken garden is ruined. The rocks are covered with silt and rubbish. I'm sure that some day I'll start again and dig them all up, but just now there's much more urgent work to do. Perhaps in the spring. . . .

THANKS, MATE!

Patricia Grace

—YEH, S'POSE I should've rung you first but I thought you was away at the sale. Anyhow I knew you wouldn't mind if I come along and picked up a few spuds.

—The cabbages? Well they're going to seed anyway. Knew they'd be no use to you.

—Yeh, yeh, thinned them out a bit for you. Got to keep thinning carrots if you want them to get a decent size, otherwise carrots go all long and skinny. Before you know where you are they all start twisting round each other and all the goodness is choked out. Same with parsnips, choked out.

—Don't want them to rot do you? Leave marrows too long in this damp weather, and they rot. See these brown patches here, gee!—that's rot. Nothing worse than rotten marrows. Nothing worse.

—For the cows? Mean t' say you're growing all that corn for animals? My opinion is people are more important. Feed the people first, then you worry about the stock. I mean: there's nothing like a bit of grass for a cow. All a cow needs is a bit of grass. . . .

—I know, I know. The cows are your livelihood, quite true. But people are still superior and that's my honest opinion.

—Them? Aw, well, I was just borrowing them. I mean: you wouldn't be using them for a while, and I got my chook pen to finish, a shed to make, and there's a bit needs doing to the house.

—Yeh, I know it's still your property. . . .

—Okay, okay! You're just the same as the rest of them round here; won't give a man a fair go. And it's not just the people that's against you, neither. It's the weather too. Never knew such a lousy place for rain. Rain, rain, all the damn time and all the animals dying, and me there milking day and night in a busted shed. Soaked to the skin, rheumatics all down one side. . . .

—So y' do, so y' do. But you all got decent sheds to milk in. And that's another thing I could do with a bit of timber for. You wouldn't happen to have. . . .

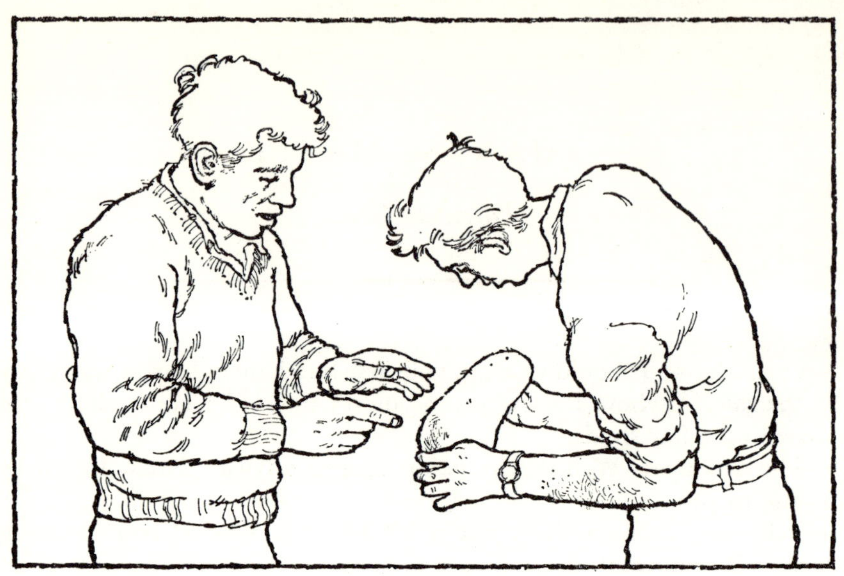

—Pay for it? Pay? Come off it, what d' you think I'm going to use for money?

—Yeh, I know you're just starting out too: but that's not the point. I reckon it's pretty lousy if you can't give a neighbour a hand-out now 'n again. Bad as that Council joker. Wouldn't even sling a bit of metal by my gate when he was doing the roads: "You gotta pay," he says. "One dollar fifty a yard." One dollar fifty? Reckon we're in the wrong business. If you can get one dollar fifty a yard for a heap of stones there's something funny somewhere. I tell you, those Council blokes look after themselves, same as the politicians. Look what you get for your butterfat. Look what you get for your pigs; not enough to make a living. But long as they're all right, buying off shares in this and shares in that, that's okay. Never mind the underdog trying to break in a bit of land with a mortgage round his neck. Never mind if we get drought all summer and rain all winter and bones aching all year round. Huh! What's it to them?

—Ah, that. Well, I can see you got plenty there—I mean, your shed's full. That miserable little bit on the truck there, that's just the overflow. . . .

—What d' you mean, it's got to last you through winter? Winter's not even here yet. You got your chain saw there, and a good stand of teatree down the gully. What more d' you want?

—Now you're getting nasty. You want to watch your step, mate.

—Okay, okay. If that's how you feel about it you can keep your firewood. Here, give us a hand and I'll chuck the bally lot down from the truck. Suppose the wife'll have to chop what we need this winter. Not much I can do with this back of mine, and you know how cold it is round these parts of a winter. And me coming in from the shed soaked through. Rheumatics all down one side. . . .

—Good for you, mate. Thanks, mate, I'll leave it on the old truck, then. Now I better get a move on. Des'll be wanting the truck back: and the wife and I thought we might as well have a trip to town while we still got it.

—Yeh: well, as I said, I'm just *borrowing* them. I mean, you're not in any hurry for them, and I just can't afford. . . .

—Yeh, good. . . .

—Yeh, I'll borrer Des's truck again sometime and bring them straight back. . . .

—Yeh, I'll make sure. . . .

—Good. Good. Soon as I've finished. . . .

—Yeh, yeh. Good-oh! I'll be off, then. . . .

—Righto! See yuh.

THE SUPER SOWERS

Trevor Meyer

WHEN FARMING BEGAN in New Zealand all you needed to do to grow grass was to chop down the standing bush, burn it in the autumn, and sow the grass seed. Then the ashes, and the fertility from centuries of decayed leaves, gave the grass a good start, and the stock thrived and produced like champions. Away like a bomb.

But wait a minute. After a few years a change set in. The good species of grass started to disappear, the green gave way to yellow tones, and in some places the stock became so sickly that the country was classed as "bush sick" because the stock wouldn't last more than one season on it.

So thousands of acres of grassland in New Zealand produced less and less; and browntop, danthonia, fern, and second growth of some bush shrubs like mako-mako, towai and five-finger, began to take control of the land again. Things were looking pretty grim in some parts of New Zealand.

That's when farmers had to think of applying fertiliser to their land: topdressing, it's called. This was a new-fangled idea . . . "How could such a little bit of fertiliser make any difference?" farmers asked. "And think of all the *money* it's going to cost," they said pessimistically.

But some reckoned that anything was worth a try, so they bought a few tons of super or slag and slung it hopefully over their paddocks and waited for results. They were marvellous: the colour came back into the grass, the better strains reappeared, and the bloom came back to the coats of the animals. Yes, it was worth it, after all.

On Puketotara up in Northland, in the far north, we faced the same problem when our bush paddocks lost their first flush of production. The cows soon told us all about it: they'd walk back up the steep hill and hang round the gate to get back into the topdressed paddocks on the open country we'd brought in from the fern. Something had to be done, so we decided to topdress.

But getting the stuff on to the paddocks wasn't going to be so easy. The cows could go down the steep ridge, but we couldn't get a sledge down, and anyway there were stumps and logs all over the place:

littered like a battlefield, so you couldn't drive a sledge round even if you got it there. So we put a big strainer in on top of the hill and swung a coil of No. 8 wire down to a stump in the middle of the paddock. Then we sledged the super, half a ton at a time, down to the strainer, split the bags into three, tied an iron hook to each bag, and sent it hissing and whining to the stump down the hill. The first few bags belted into the stump and formed a bumper for the others coming after them, thanks to gravity.

When the super was all down we scrambled down the hill, put on our sowing sacks made from grass-seed bags, filled them with super, slung another bag on to our shoulders and humped it off to the far corner of the paddock to start sowing. Bags then were twelve to the ton, mind you, not twenty as they are today. We used the logs as markers to show where we'd been, and when we used up the manure we'd carried out, we trudged back to the pile for another load. It was tough work, and between milkings of the blessed cows night and morning we managed only half a ton of topdressing a day.

Round the hills the wind often caught the super as we threw it, and blew it—*whang!*—straight back into our faces and eyes. And closing the eyes only made them sting more. I remember even now how the stuff sticks to your skin, makes your clothes so dry that they chafe; and the soreness of my arms and legs from the movement of the sowing bags rubbing on them.

But, no manure, no fertility; so on we went. We had to. There's a lot of persistence in life, just shoving on from day to day.

Then a wonderful machine came into our district: a thing called a bulldozer. It could chew tracks through the bush and bite chunks out of a hill in a way that we'd dreamed of but never thought possible. We did a bit of bush surveying and reckoned we could get a grade down to the paddock that would let us get down with the sledge.

"It's going to cost you thirty quid," said the bulldozer chap.

Thirty quid! It was a lot of money to us, but as we thought of the time and labour it would save us we reckoned it'd be worth it. So the machine went to work, and even the cows were pleased with the good road it made. Mind you, it was still a bit on the steep side, but with a chain round the runner of the sledge it didn't run on to the horses' heels going down. And there were only the empty bags and ourselves on the sledge coming up. We could put on a ton a day, a big advance, now that we didn't have to muck around with the old aerial wire. One dry year we burnt out most of the stumps and logs so we could get around with the sledge too.

Then—wonders never cease—another new machine hit the market. A truck that you filled with manure, and the whole floor wound slowly back and fed the manure out through two spinners mounted at the back. The contractor was topdressing our top paddocks one day, and had a look down our road.

"Reckon I could get down there and do most of all that with this machine," he said. "It's a bit steep, but on a dry day it should be right."

We were dubious and rubbed our chins, but he was confident so he got the job.

"It's damned steep all right," he admitted as we sat in the cab of the truck and braced ourselves against the windscreen to keep from falling forward.

But we made it all right. I tell you, it was great sitting in the truck just riding round while the super flew out behind. This was the life. . . .

"Better stand half-way up the hill with a hefty chock, Trev, in case she doesn't make it," Jim the driver called out to me.

So I grabbed a piece of old log that was handy and took up my position, but it wasn't necessary: the truck went up like a bird. It had taken only four hours to do what we had spent twelve days doing when we first started topdressing. And we didn't feel tired out, either. The only trouble was the steep hillsides; we still had to crawl round them. But at least it didn't take us so long when there was so much else waiting to be done around the farm. So, late in the spring when the weather was right, the trucks made their annual runs up and down the track, and our paddocks grew lushly. You could see them changing before your eyes: this green bloom, or sheen. And in winter, the cows sheltered snugly in the valleys while the cold winds blew on the higher country. We even got out of doing most of the steep stuff later on, when a chap with a crawler tractor mounted boxes above the tracks and a spinner at the back, and was able to cover most of what the trucks couldn't get over. But even so, there was still some we had to do by hand.

Well, the years slipped by, a war came and went, and after the war the government land development schemes hit our district. The idea was that once a few thousand acres had been grassed the Lands and Survey Department would maintain and improve it with the most modern of all topdressers, the topdressing plane. And what a sight that was! We watched the planes swoop off the ridge above our farm and scatter their loads across the country, and wondered if this wasn't the way to get our own small farm done.

So next time the planes came we had a yarn with the pilots about it.

"Yes. Get your super up here and we'll fly it on," the boss-pilot agreed, handing out smokes, fixing up the details smartly, and then driving away.

So a day or two later, out from the bulk depot come the trucks with their loads, and dump them on the end of the airstrip. A big loader hoists it into the planes, the planes shudder and sway a bit as the weight catches them, and off they go. Down the strip and over the bush into the valley where our paddocks are, then there's a trail of brownish white dust and the planes sweep in for another load. Beauty! In fifteen minutes the paddocks are all done. Fifteen minutes! And the steep parts too! It used to be a twelve-day job. And no chafed arms or thighs, either; no screwed-up smarting eyes. Nothing short of marvellous.

The sledges have almost disappeared from our farms now; and the horses, too. And the gangs of men, who used to clamber up and down and over the hills doing this dirty and tough—but oh, so necessary—job, are needed no more. But whenever I get a whiff of super as the planes fly over, my thoughts go back to the days when we used to hump the super around among the logs. I look up to the men in the clouds with gratitude. Over to you, boys. And thanks—thanks a lot.

RABBITS GALORE

Duncan Stewart

PEOPLE NOT LIVING in Otago round about that time would never believe me when I say: a goods train on its way to Dunedin from Clyde (or from a good many other parts of Otago, for that matter) could have anything up to 20,000 rabbits on board.

But believe me, it's true enough. And we were exporting rabbits by the million. In New Zealand, through the rabbit, entire fortunes were made as well as lost.

The frozen rabbit trade boom of the very early 1900s was big business, a large part of our national exports. There wasn't a railway station or a siding on the Central Otago line that didn't supply its truck or trucks of rabbits for Dunedin. And in Dunedin, these mountains of dead rabbits would be skinned and processed for freezing in readiness for the United Kingdom market.

How did this plague of rabbits begin, anyhow, in our country?

The very first rabbits were imported into New Zealand from New South Wales before 1838, but they weren't introduced to Otago and Southland until about 1850, when the first ones were let loose in the sandhills between Invercargill and Riverton, and at Queenstown, and at Rugged Ridge on the Waitaki River. These furry pioneers down South didn't do too well. So about fifteen years later, rabbits were reintroduced, and from these ones in the 'sixties began the great rabbit plagues of the later years.

Strangely enough, a number of farmers and their sons received their start on the land from the first money they got from the rabbit, but unfortunately, there were plenty of places where the bunny actually ate the runholder out of house and home.

Winter work was very hard to get in those harsh times in the early 1900s, but there was always a chance of knocking up a fair-sized cheque if you could get a block of land for trapping rabbits and rustle up the necessary equipment: a tent, unless the cocky provided you with a hut, and eighty or so steel traps with a horse to carry them. On top of that there'd be horsefeed, saddle, cooking gear, and probably a rifle. Occasionally an accordeon completed the list. Sometimes the

trapper took a dog along for company. It was mostly a lonely and arduous life, the hours long, the meals very irregular.

I lived in Middlemarch, Central Otago. In our village were representatives of three freezing companies. They had rabbit carts especially made, with wooden bars instead of floors, and frames on the sides supporting more bars. Across these bars the rabbits hung in pairs. Carting contractors were paid to travel out into the countryside ten or fourteen miles one way to collect the rabbits. This work was very trying, calling for daylight starts, very often in biting cold weather.

The rabbiter would load his horse with the eighty or more traps and set off across the tussock ridges—assuming Brer Rabbit had left any tussocks. The rabbiter, carefully picking the dry sunny spots, would start work. Swinging his grubber, he'd make a cut in the ground, about three inches wide by twelve inches long and three inches deep. Then he'd drive in a peg, set the trap, and cover it over with fine soil. It was the smell of this fresh soil which accounted for the downfall of most of the innocents abroad.

Then, if the trapper was changing his direction right or left, he would make a special cut in the ground: a marker to serve as a guide and help him find the next trap when he was making his rounds at night time. The space between traps depended on the population of the pest, though mostly it was about twelve paces.

The rabbiter wouldn't finish setting his traps too early in the day because the early-caught rabbits would be attacked by hawks. Getting back to his camp around teatime the average rabbiter lost no time in getting the frying pan into action, at the same time boiling a hefty piece of meat for next day. It's very likely, though, that the first thing he did was to have a good swig of cold tea from the billy, left over from the morning. Besides, a smoke went better with a drink of that sort.

Teatime over, the rabbiter had to start the other half of his day's work. Setting out with his four-sided tin lantern with a guttering candle inside, or in later days, a kerosene lamp, he would start to collect his catch about an hour after dark. This sounds fairly simple, but the lanterns were pretty poor quality and on a windy night would go out on the least provocation. Perhaps a sou'-wester would suddenly spring up, the horse become fidgety, the light go out. Man and beast would beat it as quickly as possible back towards home and billy tea.

Many a night my brothers and I watched the twinkle of rabbiters' lanterns, lonely and remote, far out on the ridges surrounding the small plain where we lived. Another of their time-consuming jobs at night was looking for a trap which had been dragged away by either sheep, wild cat, or rabbit.

When the saddlebags on the horse were filled with a part of the catch, the dead rabbits were emptied on the ground, gutted, legged in pairs, and left on the ground till the morning, or taken back to camp to lessen the load in the morning. After all these nocturnal activities, the trapper would possibly arrive back at home sweet home at about 10 pm or even later.

A daylight start had to be made next morning to forestall the hungry hawk which, besides picking holes in trapped rabbits, cause the rabbit to jump around until he very likely bolted, leaving his leg behind in the trap.

That morning's catch was processed, brought back to camp and added to the previous night's lot. The total then was taken to a meeting place prearranged with the carter.

When the rabbits were plentiful—yes, too plentiful—some talked of getting full-traps catches, but I think a tally of 130 from say 80 traps for both night and morning catches would be a pretty fair estimate.

When hard frosts put an end to trapping, the pea rifle came into action, and the bunny was shot for his skin—he'd have a good coat of fur in winter.

A syndicate of us used to buy 10,000 rounds of .22 short ammunition at sixpence halfpenny per 50 rounds. You could then afford to let fly at any rabbit running or sitting at 100 yards, with, of course, unpredictable results.

One year at the beginning of winter my father asked me to go with him on a visit to a rabbiter camped out in a notoriously wild and lonely gorge, about fourteen miles from civilisation. As four inches of snow on the slopes made the going tricky for the two hacks, we walked the last two miles. Abruptly, we came to a rise overlooking the tent on a little flat. And what a surprise! We were instantly assured our friend was in the best of health and spirits, for he was standing in the snow playing the bagpipes! If ever there was a replica of a Highland scene, there was one: a quick running stream close by, towering cliffs above the piper, the snow to accentuate the wild scene, the reverberating music—well now, I ask you, wouldn't it have stirred the breast of any Highland man?

And what sort of fellow anyhow was the average oldtime rabbiter?

At heart, a very generous chap who made you welcome at his caboose at any time, and you didn't get rabbit stew, either.

There were wild colonial boys among them too. With a grand flourish, they would hand over to the publican their cheques, shouting: "Let's know when it's cut, boss!" In 1920, when skins were two bob each, it was sometimes considered a deal to hand over skins in exchange for drinks.

Often my brother and I, taking the long way home from school, watched enthralled the fights in the backyard of one of the pubs. Real Donnybrooks, some of them, and after all, those rabbiters had to have some sort of compensation for the quiet days and nights on the hills.

Some of their names come tripping back. Old Jimmy Ramsay, a world wanderer; Joe Jennison, later to become mine host at Te Anau Hotel; Bill Ellis, retiring with a good wad; Joe Schriffer, who kept a big family in real comfort; and George Young, who would set 160 traps and re-set half of them, and shift the other half next day. There was Adam Wilkinson who used the rifle—choosing the .22 short ammunition—for his living. His biggest tally for one day was 127. And skins were a bob a piece then.

One rabbiter kept a stack of the classics in his camp. He knew them from cover to cover.

The gutting, or skinning of rabbits was a very smelly dirty job. You could wash your hands a dozen times, and still smell rabbits. What embarrassment a young fellow would feel when, fresh from rabbiting and dressed in his best, he'd head off to a dance. He'd be holding his girl's hand, in a dance, until sooner or later, the heat would send forth that stuffy rabbit odour.

But the girls could take it. It's surprising what a young woman will overlook if a young chap has plenty of cash to jingle in his pocket. Don't you agree?

TAMING THE TIGER COUNTRY

John McCaw

BILL BENSEBOL was a chemist in Dunedin. He had done pretty well, married a little late in life, and at the end of 1908 found himself with three young sons showing signs of becoming strong men. Bill had had a yearning to go farming all his life, and now he had the cash and the potential labour to have a go.

The North Island Main Trunk railway had just been opened, sections were available, men with sons had priority, the land was good and cheap though access was bad. This sounded right, just what was wanted: something tough and challenging, a pioneer job. So Bensebol applied for a section at Otunui, about a dozen miles from Taumarunui: 250 acres of heavy bushland, easy in contour but difficult of access. He sold his business in Dunedin and with wife and three boys aged seven, twelve and fifteen set out from Taumarunui on foot, leading a packhorse with tent, axes, and food.

The track was rough and travel slow. By the time they had reached their section, found a small clearing near a stream and pitched their camp, it was dark.

The early birds' chorus next morning, and for many more mornings, was drowned by the sound of axes and the crash of falling trees.

In two years' time enough bush had been cleared to allow twenty cows to be milked. Every third day one of the younger boys or their mother walked out over the high hill and down to the railway at Okahukura leading the packhorse loaded with two cans of cream. These were railed to Frankton and the horse reloaded with stores before beginning the four mile up-hill trek home. As often as not the cream was rotten on arrival at the factory and rejected.

Then, with no warning, disaster. Mr Bensebol died of a stroke. He was working in the bush, staggered and cried out in pain, and was gone. With him was his fourteen-year-old son, who ran home for help. Too late. The older boy hurried to Taumarunui to the police who, with neighbours' help, took in a coffin and carried out the dead man.

Mother and sons carried on. The youngest, Tom, took charge of the herd. He milked in the morning and now, every second day, rode to school in a sack on one side of the packsaddle to balance the cream can on the other. In the evening he milked again: frequently the whole herd, by himself, and by hand. No machines for the Bensebols yet.

Three more years and young Tom, returning from school, found his second brother laid out on his bed unconscious. He had been hit by a falling tree and carried back to the whare, a wooden cottage with totara-shingled roof, by his older brother John. John was already on his way to Taumarunui to fetch the doctor, who came and diagnosed bad bruising but no bones broken. Fomentations and careful exercising of the bruised legs were prescribed.

For three weeks mother and boys carried out the treatment, but the great black bruises remained, and the boy, a big strong lad, cried out in agony at the exercises. Again the doctor was called. This time he ordered the boy to hospital. Six men carried him through the bush to Okahukura to the train, from there to Hamilton, where he died under the anaesthetic, as they endeavoured to patch up a shattered pelvis and hip socket.

John and young Tom with their mother were left on a half-cleared block of 250 acres, far too small for that class of country in the days before topdressing. John was twenty, Tom was fifteen and the year 1916. John went to the war; Tom left school. John was back in 1920 but with him came a slump. Butterfat was eightpence, wool threepence.

But they were used to living off the land and they carried on. The fellow next door walked off his farm. John got it under the rehabilitation scheme. Times improved. John married. Tom and his mother farmed the home block and bought another adjacent section.

Now it was 1967. Tom had just bought a stud Hereford bull for a thousand guineas. I was admiring his purchase with him and sharing his pride in his new possession.

We started to reminisce, and he told me the story of more recent years. Mother had lived long enough to see her sons well established, and play happily with her grandchildren. John was retired now, living in Tauranga, and Tom had a fine unencumbered farm with a sealed road for access and three sons for his labour force.

Gone was the isolation. Gone too the stumps and the pig-fern. In their place sealed roads, airstrips, topdressed pastures, and pedigree stock.

But there was still the love of the rugged hills and the will to force them to give their wealth of wool and meat.

Tom loved his farm.

He came to it as a boy.

He loves it still.

He will leave it as an old man—but not until death claims him. His body, so strong to tame the virgin land, will return as dust to nourish it.

A different story in a different district and with a very different ending is that of Jim Palmer and his poor rough farm. Jim was just too young to go to the war of 1914–18; if the old Kaiser had hung on for another six months, young Jim would have been able to pull his moustache, and perhaps come back qualified for a rehab. farm, for that was Jim's one ambition—to get a farm.

So he spent a few years working on farms—fencing, shearing, scrubcutting, anything for experience and a few bob. He got plenty of the one but not much of the cash, for the 1921 slump had hit the farming business pretty hard and money was scarce. But by frugal living and long hours of slogging Jim saved up a few hundred pounds, and in 1930 found himself in possession of a couple of hundred acres of fifty-bob-an-acre scrubby yellow clay in the gumlands up North.

When they arrived at the door of the ex-Army hut that served as the homestead, his week-old wife nearly turned back to go home to Mum. But she didn't and the two of them got stuck into the job of slashing and burning the scrub, digging postholes in the cast-iron clay, sowing grass seed, broadcasting manure with the knapsack thrower, and all the other heavy manual labour that went into breaking land in those days before the tractor.

Jim had no ploughable country so he had no horses; he tramped the hills in his heavy boots and carried his gear on his broad shoulders. In two years he had a good deal of grass, enough stock to eat it and a great load of debt. Then came the Depression. Jim's careful budgeting had all to be jettisoned and a new budget prepared.

Instead of a pound for his lambs it looked like ten bob, and eightpence for his wool instead of fifteen. Even his three house-cows' bit of surplus butterfat would bring in only one-third of his estimate. How to make ends meet and provide for his wife and a family due in a few months time? Jim had never spent a shilling on himself, no beer, no baccy, no betting, so he couldn't economise there. He had no old ewes he could kill for mutton; every one was needed, too, to produce lambs. There wasn't much he could do; but he did it.

He planted a bigger vegetable garden. He skimmed the milk before using any in the house. He cancelled his paper and phone and his mail delivery. Once a month he walked to town and came home with flour and tea and sugar and precious little else. Every cheque for wool, lambs, or butterfat went straight to the mortgagor, who, to do him justice, knew little or nothing of Jim's financial plight. Jim's wife went home to Mum when the baby was due. She died in childbirth—too weakened by malnutrition and overwork to survive a bad malpresentation.

Jim, alone now with a son to work for, carried on. Fortunately his mother-in-law was able to care for the baby but Jim found that his revised budget was nowhere near the mark. His lambs were worth only six shillings and his wool but threepence halfpenny.

Too proud to ask for relief under the Mortgage Relief Act he boxed on with literally no money; every penny went in interest.

He grew a little wheat, ground it himself in a hollow stone with another stone for a pestle, and cooked it as damper. For two years he lived on skim milk, vegetables, and damper while working eighteen hours a day. Then his body gave out, some chemical process broke down, and his tissues ran to fat; his stomach distended like a famine child's. He struggled on unaided for some months, his great body hindering every step, until finally he collapsed in a coma. When his neighbours found him after two or three days he was too ill to recover.

He had fought a good fight.

And the wee boy?

A trustee company farmed the place till the boy was able to take it over in 1951, when Fate did its best to square the deal by giving him fifteen bob a pound for his wool clip.

THE STONE

Neville Peacocke

THE STONE WAS BLACK. He turned it in the late afternoon light. The colours came gleaming, first green, then yellow, then occasionally, in a sudden flash, red. He sought the red, not knowing why. He felt he had compassed the depths of all colours, all degrees of brightness, and dark.

The river was in flood. Rain had come and gone. The wetness held to the leaves, and the trees huddling into his valley brushed damp and very green. Fallen pungas buried into the earth, saturated and spongy; and grey sand above the water's mark pressed sodden. But evening was approaching, and with evening the promise of a clear night, of a fine tomorrow, and perhaps of a cloudless sky, for this was August.

Any fool knows that August is the month of the great awakening, the lambing, the new month, the month when work looms and comes crashing about one's harassed frame, when torments of a deluge of ewes with lambing troubles follow on and on and on, in endless sequence and emergency—or seemingly endless until the unexpected lull: the sun's warm light after the shower; the pause beside the river; the dismounting from the horse; the fascination of the stone, and the lights—those lights.

Why did the red in the stone attract him? Was it because he could not explain it in a black stone? Was it because it held some inner strength, some inner balm? Or perhaps the fact of real and positive colour in such an unlikely place, in such grey and green surroundings, was enough to set his mind racing?

Work eased, and the demands of seeing to things: the length of grass on the east slope; the condition of cows on the river flats; seeing if a fence was sound where it had given before; and the noise calmed till the only sound was the occasional cry, to heaven and the bush, of a distant tui, or the positive plop of a moving stone on a whispering water. Then it was that his vision became more perceptive, as if his life crept out from his body and drifted, away through the stillness and into the surrounding hills, feeling towards and sensing the essence of their colossal and evident fact. Hills made of earth, of stone, and of water. Individual pieces together forming their massive bulk. All lifting up, combining and establishing their part in unified life, forming a tremendous and magnificent voice crying "Glory!" to a great God. . . .

Was he, in existing, a vital part of all life? He could only falter on the fringe of what was perhaps profound truth.

The silence became his bridge, his catalyst, even his stethoscope. With silence he became conscious of a great life force. What mattered to him was that he was aware of, was sensitive to, this force. The awareness brought him closer to his God. He was in sympathy with the earth, the river, and the constant, familiar hills. He would cry with them; rage at the sight of an empty jam-tin, squashed and jagged, left, scornful of the fitness of things, by some picnic party; laugh with them at the sun shining on the white breasts of native kereru, a pair perched brightly together, white wedges on a horizontal branch stretched from the highest reaches of a calcified kahikatea.

Was the red in the stone symbolic of the fires of struggles in his life? Infuriating, senseless struggles against the encroaching world: the warnings of cautious trading banks that progress means regular reductions of red figures in monthly statements, greater capital equities, and fewer demands on lending authorities.

Progress! Progress towards what? Solvency? Money in a bank? Home in the city? Tarseal? Television? Tailormade recreations?

Surely such warnings are meant for less restive wills. To hell with progressive policies! Let creditors be draped with their black ribbons and filed in their columns, within their own credit margins! God gave to man his peace, to do his will in accord with the law of all things; and so to do he, in his way, intended.

There it was again, nothing wishy-washy about that red, nothing indefinite or pinkish in that deep swirling colour, nothing but depths of integrity, of love, and of rage, and—yes, of being part of, of fitting. The red had become one with the fresh smell of water, the sand, the pebbles, and the rock; the trees, the pungas and the hills; the birds and the wild. . . .

Lifting over the trees on the verge of a great central plateau, a harrier hawk rose serenely, climbing steadily until high above the bushland which spread sweeping to the south as far as the snow mountains and the great lake; merging to the west, to the north, and to the east with bush-burns and marginal farms, with clay tracks and metal roads, and there meeting the van of a creeping civilisation. Now the hawk, reaching a point above the river, circled idly, watching the man motionless beside the horse—motionless, as the sun dipped lower and lower, until the evening breeze stirred.

With it the man, moving quietly, threw the stone. He watched it bounce and fall to the dark water. He mounted the horse and rode away from the setting sun towards the east and into the approaching night.

THE BOSS WHO LIKED TURKEY EGGS

Chris Strom

THE BOSS ONCE TOLD ME when they lined up for injection in the Army it was all the big men that fainted. Now my boss was a big man. . . .

We had a great mob of turkeys running wild all over the farm. The boss had once bought half a dozen turkeys for their eggs (which are beauties, being both big and tasty), but inevitably we couldn't find all their nests, and so new crops of little turkeys would be appearing with monotonous regularity from all directions on the farm. The boss gave as many away as he could, in fact most farms in the countryside for miles around seemed to have odd descendants from our mob keen to get rid of as many as he could before the next proud mother business if only he hadn't made it so obvious that he was desperately strutting around. He could easily have gone into the turkey-selling presented him with yet another sizeable clutch.

But he'd never kill a turkey. Plenty of times people had suggested he should "do the obvious" if he wanted to restrict the size of the multitude, but he usually just looked thoughtfully at the cowshed, or the race, or the main drain, and said, "Yeah, that's a good idea," but he never put the good idea into practice. There were always so many other things to do, he reckoned.

Sure he'd give them away left, right and centre to anyone who wanted to kill the turkeys themselves. But kill them himself? Not he. The boss prided himself on being something of a he-man, often obliging us with bloodcurdling stories of impressive war exploits, not to mention his many hunting trips in the King Country in his younger days, and a display of squeamishness over lopping the head off a very tame turkey didn't quite fit the Crump-like character he'd made himself out to be.

Well, just about three months before Christmas, the boss's wife started a turkey-for-Christmas nagging campaign. A determined woman can be murder with her tongue. Once the boss had said his automatic "Yeah, that's a good idea," he was a gonner. When he brought in the usual: "There's so many things I've just got to do," she got him to itemise them, she drew up a list, and she hounded him day after day until she cornered him just before Christmas, with a day to spare on the next day and nothing planned for it.

"You can kill and clean the turkey tomorrow, and we'll put it in the neighbour's deep freeze until we want it for Christmas."

"They won't have room for a turkey," wasn't a bad reply on the spur of the moment, but she had already arranged it all, of course. The neighbour was expecting it. No, she didn't feel like going to town tomorrow; no, not even for a new hat, or a new coat, or a new dress, or a new outfit. No, she didn't want to go to the beach. Just that turkey for Christmas, and she'd be very happy, thank you.

Of course, if he didn't want to kill a turkey then the neighbour would be only too happy—all right, there was no need to swear, she was sure he could kill a turkey, certainly he just lacked the time. But he'd have all the time he wanted—and more—in fact all day—tomorrow.

I reckon it must be terrible to be a boxer punched into just about helplessness and seeing the knockout punch on its way towards you knowing there was nothing you could do to stop it. The poor old boss. You'd have thought he was the one who was going to have his head chopped off. He mooched off to bed early, mumbling something about the weather could break, and it might rain all day tomorrow. Well, they say hope springs eternal. . . .

But next day broke beautiful and fine. The boss looked terrible. I got the impression that he hadn't had much sleep during the night, and all through milking he was grumbling to himself and cursing the cows. Right through breakfast, his wife chatted and clattered gaily around while the boss sat in surly silence, never lifting his eyes from his tucker. Then he offered a bit sheepishly "to give a hand with the dishes"; and when his wife's jaw had ceased sagging with amazement, she thanked him, vanished out the door, returned with the axe, thrust it into his hand and banished him out to look for—and kill—"the juiciest, plumpest turkey on the farm".

Walking with him, I felt like a warder escorting a condemned man from the death cell to the scaffold. We must have looked as cheerful as a pair of undertakers.

The turkeys were easy to catch at first, but then the boss had doubts about the size of first one, then another. He was forever seeing a bigger one in the next paddock. When a sizable one took a nip out of my left hand, a look of relief came onto his face.

"We must go and get that bandaged up," he said, and advised me to drop the turkey. But I thought his wife might like to see this monster bird. The boss gave me a funny look, and we plodded dejectedly back to the house.

Sure enough the missus was delighted. She held the turkey while the boss wound yards of bandages over the scratch on my hand, the boss saying all the time that he should be had up for letting her hold such a ferocious bird and that all turkeys were as wild as hawks. Finally the first aid lesson was over. The boss's wife was now standing right by the chopping block.

"Will I hold the turkey or do you want to hold it yourself?" she asked.

The boss suddenly thought the axe was too blunt, and went off to find a file to sharpen it. He was gone so long that his wife handed the bird to me, and went off to find him. But first she had to rush inside to put some sticking plaster right where she had got a really nasty bite in the crook of her arm. She finally rounded up the husband, who emerged from the depths of a shed, blinking like a surprised owl, a hammer and an old World War I bayonet in his hand. His wife had seen birds killed in that way. You stretch their necks out on the block, she said, put the blade on their necks, and hit the blade with the hammer.

The boss thought this sounded like a good idea too. So I held the turkey's body, the missus held its head, and the boss placed the bayonet on its neck.

He swung the hammer, closed his eyes—missed, and hit his hand. With a yell he dropped the bayonet, his wife let go the bird's head which, of course, promptly turned and bit me on the right hand, so I let go the bird. The turkey and boss's wife both went berserk. She grabbed the bayonet, and went lunging after the partially fenced-in bird like a Maori Battalion soldier, shrilling to the boss: *"You closed your eyes! I saw you! You big sissy!"* and other unrepeatable phrases. The boss wisely disappeared in the direction of the neighbour's farm. . . .

About an hour later I heard a shotgun go off. There were about five shots in all, over a fairly short space of time. Neither the missus nor I dare to move from the kitchen, at least not until I knew who was doing the shooting.

Sure enough, the boss himself soon strode through the gate, the neighbour's gun in one hand and a half-shot-to-bits turkey in the other, and a smile of pure misery on his face. A gun was the only thing to use against such ferocious birds, he said.

Well, I dug about half a thousand pellets out of the turkey, and we found half-a-dozen more the hard way at Christmas dinner. His wife seemed to forgive him, after about three or four days.

And it's a funny thing, the boss can usually see a joke. But the neighbour and the boss were yarning not long after, and the neighbour remarked about some executioner who'd just written up his life story.

"They'll be wanting to publish your life story next," he says to the boss.

"What for?"

"The day you executed the turkey," says the neighbour.

But the boss couldn't see the joke.

THE HEIFER

Mary Gillies

SPRING WAS LATE that year. The paddocks were swept bare by incessant cold westerlies and there was no sun or rain for weeks. The grey, dull days followed one another dourly, chased by a bitter, sour wind which curdled the cheerfulness of man and beast. Although the plum tree blossomed and the daffodils bloomed to the calendar, the grass lay dormant, waiting for warm showers and sunshine to set it growing. The big leaves of the puka tree by the cowshed turned their backs to the west and rattled in discomfort.

The day the heifer was due to calve, the wind died.

At first there was a strange stillness; even the birds were hushed, withholding their songs until they were sure they would not be hurled back in their throats. The plum blossom floated gently to the lawn; the daffodils lifted their battered heads; and leaves on the puka lay shining, right side up, and were silent. Then, from the north drifted a warm, soft breeze, and by evening a misty rain was falling.

The heifer had been uneasy all day, following me about the paddock and trying to thrust her head between the bars of the gate when I left her.

By nightfall she had started to calve, and the earth under the old totara was scuffed with her straining. All the signs pointed to a normal birth, and there was no reason for me to stay with her, except that she wanted my company. Between birth pangs, she stood quietly, chewing her cud while I stroked her neck and spoke reassuringly to her. When I tried to leave, she voiced her agitation, so I pulled my coat about me, crouched under the totara, and prepared to wait.

The other animals in the paddock were aloof and uninterested, vague humps in the gloom. I could hear some of them picking at the short grass, and others breathing heavily as they lay, somnolent and contented, digesting their evening hay. A truck backfired as it careered down the sideroad past the farm, and I heard the sickening skid of braked wheels as it almost failed to take the corner at the foot of the hill.

ERIC HEATH

The birth, though normal, was slow.

The heifer lay stretched on her flank, grunting as she strained. The muzzle of the calf, tongue protruding sideways between tight jaws, lay above its soft, tender hooves. At last, after a long, stern struggle the whole head appeared, as ugly as a gargoyle. For all my vigil, I didn't witness the final, triumphant moment of birth. As the heifer lay down for her last effort, the moon broke through the light clouds for a few moments, and threw across the sky for my delight, the pale, pure ghost of a rainbow. Entranced, I forgot my calving cow, and stared until the moon was shrouded again, and the faery arc was gone.

The harsh, unlovely hour of travail was over and already forgotten by the heifer as her rough, strong tongue worked over the calf's body. Quickly the tongue cleared the birth-coating from nostrils and mouth, and the calf drew its first gasping breath. A bubble ballooned from one nostril as it breathed out. The tongue worked on tirelessly, lovingly, around the eyes, over the ears and along the neck and body. With its first breath, the calf began to struggle, every muscle and nerve directed towards one end: sustenance. Strength snowballed with each seemingly aimless, unco-ordinated movement until, in an incredibly short time, it attempted to stand, flopped to the ground, and tried again and again until, inevitably, it succeeded.

When I left them the calf was sucking greedily and noisily and the heifer, head turned, was licking its hind quarters, uttering low sounds of love and pride.

I lifted my face to the rain, which fell like a soothing lotion on my wind-chafed skin. All my little world was shining as I walked home across the paddocks.

HOME SWEET HOME ON THE FARM

Alice Dwyer

POOR PEOPLE? Started off seven cows, one horse, one dog, chickens and cats. Open door and cupboards full. And home always full of people. There were twelve in our family, and a kind and good father and mother, plus at least a couple of dozen for holidays, and visitors always on Sunday; friends and neighbours any day. No Social Security. No Child Allowance.

Sunday our day of rest, but cows to be milked, separating to be done, and butter to make. Feeding calves, pigs, dogs, cats, chickens, and family. Breakfast: porridge, fried veg. and meat, and mother's home-made yeast buns. Hymns around the harmonium, and wanders over the hills or a row in the dinghy. Always home-made butter, cakes and fruit salad for tea, and meat and vegetable salads and cream.

Mailday three times a week. Three and a half miles walk or with horse and cart. Dancing till 3 am. Tennis, church, rowing, swimming, sewing, singing, picnics, parties and games: cards, draughts, five hundred. Always more letters to write. Drawing, painting, hair trimming. Gardening, chopping wood. Collecting mushrooms, pine cones; blackberry picking, haymaking, gumdigging. Institute. Bring-and-buys.

Darning socks, knitting, fancy work, cooking, housework, cobbing corn. Storing and bottling and jam-making. Grass thrashing and veg. and seed storing. Fishing, eeling, spearing, and rabbiting. Listening to the birds singing. Flowers. Reading the *Weekly News*. Farmers. Fertilisers. *Herald*. Sunday at home. Leisure hours and volumes of books, and school prizes, and old photographs.

Yes! The good Lord *did* provide.

Trees

THE PINE TREE

Mary Gillies

THE OLD PINE tree is dying. On its south'ard side the crusty bark is smeared with a pale green scum of lichen. One of its twin trunks is already dead, a desolate skeleton of bare, windbroken branches. The other trunk still has a few living limbs, and the sparrows build their loose untidy nests among the thinning clumps of pine-needles. After a gale, I sometimes find nests on the ground with eggs, or dead unfledged nestlings, inside.

Bodger the dog tells me that a possum skulks out the daylight hours in what shelter the tree still offers. I cannot see him, but Bodger's seldom mistaken on such matters. Some of the lower branches of the dead trunk were sawn off years ago, and the protruding end of each one is tenanted by a minah family. From the cowshed I watch them coming and going, and in the nesting season I wonder at the amount of hay and litter carried into these hollow butts.

A clear spring bubbles to the surface almost at the roots of the old tree. I dream of damming a little pool there, and growing native swamp plants about it, beneath kowhai, cabbage tree, and flax. . . .

But the pine must be felled first, and the wilderness of self-sown barberry growing around it cleared. Under the roots on the higher dry ground there's a cleverly concealed entrance to what was once an outsize wasps' nest. Before it was discovered and dealt with, the house was never entirely free from wasps, even in winter, when on warm days one or two would sleepily sun themselves on the windowsills, or crawl into the jam cupboard.

There's a comfortable natural seat at the foot of the tree where I sometimes sit. It's a good spot for thinking, for looking and, I suppose, for time-wasting. I sit there and prise off a lump of bark. A brown, creamy striped earwig, pincers at the ready, freezes a second before scuttling for shelter; and slaters, some a wondrous purple, conspicuous among their grey brothers, scurry for dear life from the flooding light. Spiders, less concerned, unhurriedly lose themselves under the creviced bark.

Idly, thoughtlessly, I probe deeper, scratching away at the earth mould between bark and trunk.

When I uncover an ant's nest the inhabitants rush out and pour over the ground, but without panic. There seems to be some communication between them, as if their leaders are organising a prearranged emergency precautions plan. High in the tree, safe from my prying eyes and fingers, a cicada cracks the silence. A moth settles on a drooping flower-head of willow weed; a pair of white butterflies hover over the watercress below the spring. Long before I see him, I hear an approaching bumblebee, noisy as a helicopter. He circles the tree, briefly investigates my hair, and goes off about his business.

Evilly, gracefully, a hawk flies low over the swamp, wheels, and comes back again. I wonder what movement has attracted him and, watching intently, I see the long grass quiver as if some small animal is forcing a passage through it. I wait, ready to scare the hawk should he swoop too low.

Then I see a narrow, ginger-and-honey striped tail tip in the swaying grass, and I call encouragement to Thomasina, one of the cats. The hawk flies off, the tail-tip vanishes, and two pointed ears and a white smutched nose appear above the grass as Thomasina raises herself upright to see me. She gives a wail of pleasure, drops her head, and

re-erects the tail-tip, which moves steadily towards me like a periscope through a grass-green sea.

When she emerges, climbs the bank and settles on my knee, she purrs in self-congratulation and delight as her head butts my chin. If cats have a memory, she can't have forgotten the fearful adventure which sent her, scarred and bleeding, to seek refuge high in the pine tree for two nights and a day. I was without hope when I found her on the morning of the second day, and I never knew what terror drove her there.

The old tree has had its day. It's an eyesore now, and a danger. Common-sense demands its removal.

Why, then, am I so reluctant to say the word that will set the chain-saw whining?

OUR FRIEND THE WEEPING WILLOW

Patricia Clemett

MUCH WATER has passed down the creek since I last saw my own special weeping willow tree, but I never visualise our old homestead at Karamea, on the West Coast, without it. There it stood, a giant of its species, its long, weeping, sweeping branches almost touching the ground and a close green world inside.

My first memories of this tree were not particularly happy ones; it started life in a rather unusual way. Alongside my father's place at the table was a handy switch, which served as a deterrent to any misbehaviour at meal times. I was one of the younger members of the family, and not much notice was taken when I showed a preference to sit close to dad's right hand at meals. One day, when I had deliberately misbehaved, the rest of the family saw that dad had great difficulty in wielding the stick at such close quarters, and after that I found that, if I wasn't first at the table, I would miss my seat.

One day a rebellious member of the family took the switch and threw it, like a spear, out into the backyard, where it embedded itself in the ground. Weeks later mother noticed that the stick had shoots on it, so our one-time corrector was left to take its chance of survival. Survive and thrive it certainly did. Yes, the tree really did grow from dad's switch.

Weeping willows are usually slower growing than the ordinary willow but this one gave the lie to that belief, and long before I left childhood behind me many a happy tea-party I held with my friends under its branches. A piece of rope tied between two branches made a handy clothesline. And each year, when Anzac Day came round and schoolchildren were called on to make wreaths, our tree was in great demand, for what better than its soft, pliant branches to form a frame for the flowers? And dad, a veteran of the Boer War and the First World War, with medals polished, would march proudly in the parade, happy in the knowledge that he had helped in the making of those tributes to the fallen.

Mother's greatest pleasure from the tree was when a schoolteacher friend cut many fine branches and wove a much needed clothes-basket and gave it to her, with the advice to leave it out in the sun and wind to dry out properly. She forgot to tell mother what to do if it rained, and to this day we delight in telling the story of how mother's clothes-basket "grew". What if the basket was heavy and the clothes became a bit stained from the improperly treated branches? Not to worry. Mother simply lined it with fresh newspapers and would never part with it. The willow tree made a handy repair shop in our backyard, too, and gave a constant supply of material to renew any part which became broken, and if the basket showed any signs of sprouting the shoots were quickly removed.

My two older brothers, usually the best of friends, would become the most bitter enemies when one would take the other's favourite cow at milking time. Differences weren't fought out at the milking shed—oh, no! That would upset the herd and reduce the milk and butterfat production. So it would be with tempers almost at boiling point that Barney and Laurie would reach the homestead, kick off their gumboots and, by mutual agreement, make for the amphitheatre close to the trunk of the tree.

"Don't let me see you boys fighting when I get home," mother used to admonish them before they left the shed.

She never did. The drooping branches made a perfect screen from all but a highly privileged audience: my brother Jack and I perched precariously above the battlefield.

The light evening breeze would rustle the branches, and the old tree would sway and sigh as if in protest at this unseemly display of human emotions. After a fight, a strong branch for them both to lean on, the victor to regain his breath and the vanquished to nurse a bleeding nose. Then, the mopping and cleaning up done, they would each grab a hefty branch of willow-wood, a piece of thick twine and a lump of meat, and head for a quiet evening's bobbing for eels in the nearby creek.

Our tree was remarkable. It seemed to defy the laws of Nature, and I can't remember the time when it was entirely bare. Perhaps the lovely mild winters in Karamea accounted for this, but all the same we looked forward every year to seeing it in all its spring glory; and each autumn the first sign of yellowing leaves always brought a hint of sadness to our hearts. You may think that was carrying sentiment a bit too far—but then you didn't have to spend hours raking, scraping, sweeping, and burning the fallen leaves.

In the spring we never saw birds nesting among the branches. With so much activity around the willow, no bird even had time to lay an egg, let alone build a nest there. Never was a tree so used, abused, and mutilated. But never was a tree so loved and appreciated. I think our love and affection helped it recover from some of the demands made upon it. It continued to thrive, it shared our secrets and, in one instance, it hid our crimes.

My sister Esmé had acquired an ardent but unwelcome swain. One Sunday he arrived with a beautiful bunch of violets for her. She thanked him warmly, but when she went to put the flowers in a vase a slip of paper caught her eye. A billet-doux? No—just the first part, a bill—"1/6d. please, Esmé."

Her blush of embarrassment turned to one of annoyance as she paid up and told him just what to do with the rest of his violets. The advice fell on deaf ears, and the following two Sundays brought a repeat of this, with his violets and his bill for 1s. 6d. Those Depression days of the 1930s, when every penny counted: I ask you—who could afford to buy flowers every week? Certainly not us. Esmé and I decided it must be stopped.

So one night two furtive figures slipped into the admirer's garden, removed enough of the plants to start a little patch of our own, and did enough damage to the rest of his violets to retard their growth until next spring. Arriving home, panic gripped us. What would mother and dad say? Well, we weren't going to waste those plants, so on the far side of the tree, where the branches touched the ground, we dug a trench, transferred the violet plants from the sugarbag into the hurriedly prepared plot, then allowed the branches to fall back into place to hide the dastardly deed.

There they remained unnoticed, until next spring, when Esmé and I in great surprise discovered a beautiful patch of violets blooming beneath the willow tree. How they came to be there remained a mystery—that is, until we were both grown up and too old to be chastised with a branch from the same tree which had conspired with us.

Our parents had always taught us to be honest, but this crime paid off when Jamie the swain started bringing violets again the next year. My sister was able to show him a whole vase of violets and say, "No thanks, Jamie. Someone else brings me flowers now, and I don't have to pay."

I think our tree must have been the most photographed tree in the district. As a background for photos of anyone and everyone, it was well patronised. I remember how old dad, when he was shown some photos of my eldest sister taken in all her wedding finery, gazed intently, admiring them. We were all quite touched, as he was always so reticent with any praise or display of emotion where the family was concerned. Was he softening up in his old age? We needn't have worried.

Later that afternoon we heard him on the 'phone:

"Hello, Bill! I want you to come over and see some photos. What's that? Of the willow tree, of course. Yeah—Freddy Williams took them the day Nellie got married."

When second sister Eileen was married a year or so later, she posed beside a rambling-rose bush.

Even our old dog Tip loved this tree. It provided shelter from the rain and shade from the sun, and he preferred the trunk of the willow tree to any fence post on the farm. I have known old Tip to travel past dozens of posts just to show his preference. And when he eventually passed on, we thought Tip would always have it as a living tombstone to mark his last resting-place. But this was not to be.

Every year dad used to kill a pig for bacon. We didn't have to rig up a gallows or bother with block and tackle and so on. A rope tied to the pig's back legs, a hefty piece of willow as a spreader, and the carcase was hauled up and hung from a branch and we had our very own "gallows tree".

And there was another use, too, when the whitebait were running. In those days, to form a bow for a whitebait net we usually used supplejack, but in an emergency two branches from the willow tree, lashed together with flax, served very well.

In later years, with nieces and nephews spending holidays at the farm, miniature Tarzans would swing and shout from the branches. Tree-houses took proud but precarious shape in the higher branches, and the willow gave a constant supply of arrows for budding Robin Hoods. There are always Robin Hoods in every generation, and this tree had supplied generously three generations; and how many more to come?

THE ORANGE TREE

Melba Harris

I LIVED IN a small village which could boast only a few shops stocked with uninteresting everyday necessities. No fruiterers helped add colour to the group of drab shop frontages. I hadn't seen an orange in all of my first ten years.

Then, one day an uncle arrived from the far north. He brought with him a case of golden oranges. When the lid was levered up, I gazed in delight. Oh, the colour of those wonderful oranges—the sweet, pungent scent! I wanted to spin them out over weeks and weeks.

"They won't last very long," said my mother. "But suppose we plant some pips and try to grow our own?"

"Better than that," said my uncle, "I'll send you a small tree."

True to his word, in due time the little orange tree arrived. Our red brick two-storey house sat square and solid on a knoll with the Sugar Loaf rising up behind. Our view from the front terrace ran down a long narrow valley with a glimpse of the sea beyond. My father said the tree would be happiest leaning against one of the cobblestone pillars supporting the roof of the terrace.

So I planted it most carefully in deep loam, spreading out its roots and patting down a blanket of soil. For company I scattered small sun-warmed granite rocks and clods of moist earth that had the roots of wild white violets still clinging to them.

And the orange tree grew as I grew. Daily I measured myself beside it. At first I looked down on it. Later I looked up—up into its dark glossy branches. Excitement indeed when I spied the first white waxen buds like pearl drops! I watched them open to show an inner beauty, and the fragrance of their perfume drifted through our home.

Then Gaston came. He came up out of the sea one morning. His shaggy, iron grey hair was matted with dried salt spray, his grey eyes shrunken and haunted. A hanging shred of shirt exposed his brown tattooed chest, and his denims were stiff with salt-caked ridges. He limped badly, and one bare foot was roughly bandaged.

"Me zhip," he half-whispered in a choked voice. "Me zhip and me men—all broken on ze cruel rocks—verree bad storm—drove us in.

ERIC HEATH

Me—big wave rolled me over and over—threw me up the beach. I reached out me hands and gripped some grass, dragged meself into ze bushes. I lie zere all ze day—two nights—now I come up—up—up—"

My father grabbed him as he appeared to faint, and half-carried him round to a shearers' cabin at the back of the house. There my parents nursed him back to health. Gaston had lost everything, so my father asked him to stay on and take over the sheep that grazed on the Sugar Loaf.

He soon adapted himself to his new way of life, and proved invaluable. The old sailor was much interested in my orange tree, which had flourished, bushing out thickly, hiding the cobblestone pillar and almost occupying one end of the terrace.

"I work in ze orange groves in south of France, many years," said Gaston one day. "Some house have one tree against a post, like zis, but all trunk—no branches—and all spread out at top, like umbrella" (spreading wide his arms) "and fruit, big crop, grows all over the roof. You like me try?"

So I left my tree to Gaston's care.

In no time, after being stripped of its branches, it thrust skywards till it brushed against my window sill and sprawled its branches widely over the roof and beyond. The square brick house swelled into a carnival of blossom. My orange tree, true to Gaston's prophecy, produced almost inexhaustible supplies of superb fruit, and an almost round-the-year smother of blossom. How beautiful it was by moonlight, too. . . .

Throughout the long summer, the humming of the bees among the blossom . . . and how we enjoyed the orange-blossom honey on our breakfast table.

The years moved on, bringing many vacant chairs in our household: but my tree and I had the deepest roots of all. Then the time came, at last, when I realised that I too must move—to a Home where I could be cared for. I had to face the parting from the house where I had spent all my life. It was such a gracious old place—always so gentle and welcoming: and now it was goodbye.

"But how can I leave my orange tree?" I cried, almost despairingly.

"Take it with you!" said a friend brightly.

And so I took my orange tree with me to the Home. No varnish or stain mars its natural beauty. It is a table now.

As I write this, I move my fingertips caressingly round the edge of the oval table, satin-smooth to the touch, and golden-brown like an acorn. It is always with me here, beside my wheelchair.

THE PEAR TREE

Lorraine Parsonson

DOWN THE YEARS we've had our unexpected visitors too, or intruders, even though until recently we've had no roads in our part of Pelorus Sound, in Marlborough.

My Grandmother told me of the time when, as a bride of sixteen, she too was living in the Sounds, at Elie Bay. She happened to be standing on the shore gazing at the sea—not an unusual habit for us Sounds folk. Wishful thinking, perhaps? Whenever we got to our front gate, which is inevitably on the shore, we look out and wonder: Could anyone be coming today?

On this particular day, there was somebody for Grandmother. Around Separation Point came a ship's lifeboat with a solitary figure in it, rowing steadily towards her. As the boat touched the shore the man stood up—all six foot four of him. He was handsome, with dark flashing eyes A red handkerchief covered his head, and yes—he had gold earrings. Surely a pirate!

By the time Grandfather had joined them, the pirate was asked up for breakfast. My memories of Grandmother are when she was an old lady, and when I pestered her with questions the answer was always: "Ask no questions, and you'll be told no lies." Presumably she felt the same at sixteen, for no questions were ever asked of Pirate Jack—not even his surname. It was thought that he was a deserter from one of the sailing ships frequently anchoring at the mouth of the Sound to take on water.

Pirate Jack was there to stay. A small hut was built for him near the house, and he filled in his time doing odd jobs round the homestead, garden and farm.

At this time a timber mill stood at neighbouring Crail Bay. On winter evenings Jack would launch his boat and row there. He was blessed with a glorious singing voice, and would entertain the mill-hands by the hour. They would reward him with rum and tobacco.

The years sped on. Pirate Jack became an old man.

One day he came to the house complaining of not feeling well. Grandmother suggested a drink of water, perhaps as an antidote to the rum he'd had the night before, and as she offered it to him he fell dead at her feet.

There were no telephones then, in fact no communication with the outside world at all. Grandfather and Grandmother half-dragged, half-carried him about fifty yards away. There they dug his grave and laid him gently in, red handkerchief, gold earrings and all. They filled the grave in, said the Lord's Prayer over it, and on the new earth planted a tiny pear tree.

Now, almost a hundred years later, in spring the tree is covered with a soft white cloud of blossom, fragrant and sweet, as the bees discover as they fly busily about gathering the honey. In autumn it is laden with golden fruit, enough for us and the birds as well. In winter it holds up its branches, stark and bare against the sky.

Dangers and Joys

"HELP! I'M LOST. WHAT NOW?"

Tony Nolan

WHENEVER I READ in the paper, or hear on the news, that somebody's lost in the bush, I give a bit of a shiver, and I think: "Thank goodness it isn't me."

And I imagine myself out there in his place . . . lonely . . . night coming on . . . trees all the same, like walls, crowding around, and moving as you move so you seem to be getting nowhere. And there's noises, things that brush against you in the dark, queer shapes that seem to move, and eyes that stare at you out of the shadows.

And I wonder whether this chap felt like that when somebody else

got lost—or whether he didn't give it a thought, because it couldn't happen to him anyway. Perhaps he never went near the bush . . . and then a cow got through the fence, or a cobber talked him into a shooting trip, or he heard about those big fish up the river. Or maybe he was a kid who didn't know better, or just a bloke who stopped his car to look for a spot for a picnic or something.

I know he'll be pretty scared, and I wonder how he's coping with his fear. Maybe it's a bit silly, but I try to send him little thought-messages, saying, "Fight it, boy. Fight it." Because I know, that if he can only beat his fear he's got a good chance of seeing home again.

Fear's the biggest killer of the lot. Of course there's good fear, the sort that warns us and stops us doing silly things. The sort I mean is bad fear: panic, silly worry fear, stupid superstitious fear that should have gone out with the goblins.

Take the man who goes missing and the searchers find him lying in the river. The coroner says, "Drowned, while trying to find his way home." Well, perhaps he did drown, but what killed him in the first place was fear. That bloke in his normal state would never have risked a crossing at such a crazy spot. But all het up with fear of being lost, and the disgrace, and the loneliness, and what his folks would go through if he didn't turn up—in he went and . . . bingo!

Take the girl who gets separated from her schoolmates, the usual story of getting left behind or trying to catch up with someone in front. She's pretty game and does all the right things: tries to find her way back, tries to remember things she passed. When it gets dark she makes a bed of fern under a bit of a log and lies down planning what she's going to do in the morning. Then the noises start, and the loneliness gets at her, then a pig snorts or a possum scuffles, and blindly she rushes off into the dark. The verdict when they find her is "Exposure." Exposure? That poor kid died of fright.

You've got to beat fear, and to beat it you've got to think about it beforehand, and to know what's likely to frighten you, or worry you, and how darn silly it all is.

For instance, just after Easter I was camping by myself up the Mokihinui, a big river on the West Coast. Now I'm supposed to be an adult, but it was a lonely sort of place and I was tired out—in that state where you begin to imagine things. As I settled down for the night, three stags started to roar. *Rrrrrrrr-yeeow-rrrrr*. All around me and pretty close. Then a woodhen let fly next to my ear. *Kooweeee. Kooweeee. Kooweeee*. So loud that it didn't sound like a woodhen, the grunt part nearly rattled my billy. Then an owl joined in. *M'rrrrreeeeek*. Then, just as my nerves were settling down—I was sure those stags were closing in on me, growling like lions—a possum let out a screech that set my hair on end. *Aaaaaaaaaaaark-ha-ha-ha.* For all the world like the cracked voice of a loony old man.

"Crikey," I thought, switching on my torch in a hurry, "it's just like sleeping in a zoo."

Then I yelled back at them, and they shut up for a while. But the point is that I knew what the noises were, and I knew that if I yelled the silly animals would be a darned sight more scared than I was.

But what if I had been a lost kid? It made me think. It's time that a lot of other people, and their children, knew something about the bush inhabitants and the noises they make, particularly at night.

Another thing: lots of children are still scared stiff of wetas, moths, and other bugs. I don't like them crawling on me either. But I've never heard of anyone being bitten to death by them. The kids should be taught that.

One thing that worries lost people as much as anything is fear of starvation. Some blokes really panic when they miss one meal. Well, there's plenty of folk lasted a fortnight or more on nothing but fresh air and hope and water. I've done a seven-day starve myself. There was a Chinese seaman during the war who survived something like a hundred and fourteen days. And there's plenty of people in the world today who do hard manual labour on what we'd give to the canary.

The first three days are the worst. But hunger isn't starvation. If you can't still do a good day's hike after a week without fodder then it isn't your stomach that's at fault.

Exposure's another thing I should mention. I've known big strong men to panic at the thought of one night out in the bush without a bed. They seem to think that exposure will get them. Well, exposure is caused by a combination of things: wind, rain, low temperatures, and low morale. Beat one, and you'll beat the lot.

Wind is the worst. It sucks the warmth out of your body like a man emptying a beer glass on a hot day. I've seen blokes down the Antarctic walking around outside with the temperature at forty below zero and no gloves, and clothes just like you'd wear in the bush in New Zealand. Then comes a puff of wind, and it's head for the great indoors, pretty fast.

Maybe you can't always get out of the rain—though there's not many patches of bush that haven't an overhanging bank, or a rock or a log you can crawl under—but if you can't get yourself out of the worst of the wind then you're not trying very hard.

All you need is a bed of fern leaves, bracken or beech twigs under an overhanging log with more branches or wood on the upwind side; or a wind-shelter log with a lean-to of branches and ferns, or even a big heap of punga leaves or bracken you can crawl into. Naturally you'll shiver a bit in your wet clothes and you may have to do exercises from time to time, and perhaps you'll have to sing or talk to yourself to keep your pecker up. But in the morning you'll be alive and kicking. For centuries our ancestors did without tents and sleeping bags, and they survived or we wouldn't be here.

I've been on a few rescues and I've seen lost souls who seemed to be suffering from exposure. Of course we treated them for that, but from the way they perked up within a few hours it was pretty obvious that their main trouble was loneliness, tiredness, and worry.

It's not the bush you've got to beat, it's yourself.

Hunger gets your morale unless you watch it. Yet I don't know that it helps much to go scouring the bush for things to eat. Of course, if you're absolutely stuck by a river, or you've wrecked your legs or feet and can't travel, then you might as well spend the time looking for berries, eels, woodhens and other potential tucker. In any case, it pays to know what's edible in the bush, and how to make snares and things. If you have a rifle a good feed of possum or goat or deer is a great morale-builder. But still, in the main, I think the first day or so is best spent finding the way home while you're still in tiptop condition.

How to find your way home depends on the circumstances. There's not much I can tell you except a few basic rules.

First, as I've said, don't panic. Try to get rid of fear: beat it out of your head—with your fists, if necessary. And chase it away every time it creeps up on you. Tell it you know it's just a humbug, trying to confuse you and make you do all sorts of crazy things. Fear of loneliness, fear of darkness, fear of hunger and exposure, fear of what your cobbers will say, or how your folks will be worrying—let your folks worry: that's their job. *Your* job is to stay alive, and get yourself found again.

The next thing you have to do is to sit down and think over how, when, and why you did get lost. That should give you a good lead to your plan. It's no use running blindly around like a hen with its head chopped off. You've got to have a plan. Apart from anything else, working out a plan will give you a bit of confidence, steady you down, help you to remember things, because then you're someone doing something definite.

When you've decided what you're going to do, do it carefully. Take it gently. Don't try to rush. I'm willing to bet that a lot of the people who never came back fell over cliffs, or broke legs, or wore themselves out, or got washed away in rivers, all through trying to rush things. The old enemy fear again.

Generally it's better to try to find your way back, rather than forward. Even if you're on your way home it's usually worth going back a bit to look for the track. And better to stick to the long way and not take short cuts. As you work your way slowly back, stop, and try to remember things you saw on the way in. Things such as a twisted tree, a branch you had to crawl under, a sticky patch of supplejack, an outcrop of rock you may have noticed. Maybe you'll be lucky enough to see a footprint at a muddy bit. That way it might help you to get things straightened out.

Sometimes it might pay to climb up a hillside a bit, to look at the country generally. Or climb a tree on the hillside. But only if you're sure both you and the tree are in a fit condition: a broken leg won't help you much in getting home.

In the long run you may have to do the time-honoured thing and follow water. Creeks run into streams, and streams run into rivers, and rivers run out to settled land or the coast. But take it easy: don't drown yourself or break your neck just when you're getting somewhere. Better a long slow climb over a bluff than a river-crossing you can't manage.

And if you do have to spend another night out, stop early enough to make yourself comfortable. Even if it does mean that search parties will be called out. People don't mind searching if they think a bloke is trying to do the right thing. And incidentally, a few arrows scratched in the ground, or made with sticks, obvious footprints, and things like that, are a big help to them. They may be coming down from behind you, or across country.

If somebody knows you're overdue, then sooner or later the searchers will come out. They'd get a kick out of finding you, so think out every way you can help them to do it. Nowadays it's quite usual to call in a plane or two—so think how you can help them too, by displaying something which will be conspicuous from the air.

Of course, the big thing is not to get lost in the first place. . . .

Accidents can happen, but probably most times they can be avoided. The usual causes are separating from the others who know the way (often their fault as much as yours); making vague plans with your cobbers—you know the sort of thing: "You take this spur and I'll take the next and I'll wait for you at the top," when, in fact, the spurs don't meet at the top. Or when one bloke turns back and the other poor Charlie waits patiently until it's dark and then loses himself on the way down and finishes up miles away.

Being in too big a hurry. I suppose that's the worst of the lot . . . in such a rush that you don't take the elementary precautions of watching where you're going and stopping from time to time to have a good long hard look back—it's surprising how different things can look when you're going the other way. That look back from time to time is pretty important.

Too much of a hurry. Chasing after a deer, trying to catch up with a cobber, trying to keep ahead of a cobber, trying to get somewhere before dark, trying a short-cut home. And not looking back from time to time and not remembering those little landmarks and forgetting to put a sign of some sort at the junctions of tracks and creeks and ridges.

You know, sometimes the bloke who gets into trouble that way is the one you'd least expect. The farmer, who does a lot of hunting up in the "back paddock". The bushman. The possum trapper. Perhaps it's because he gets so good in his own country that he thinks he'll be right anywhere else. But he forgets why he can handle his own country—I think it's because he learned all about it, heard all about it, from his neighbours, or as a kid, or at the local pub, just hearing people talk about it. And, when he went into it for the first few times he took a good look at things around him, because it was his territory, and because he'd probably be going again, and because he'd be wanting to talk about it.

That whole process has to start again when you're thinking of going into country that's new to you. Getting information, looking at maps, finding out about the tracks and the local landmarks, finding out the best way up creeks and what's on both sides of the place you have in mind. Local folk can keep you out of a lot of trouble—make it much easier. And when you actually get on your way you take it gently for a bit, having a good look around as you go, so that you can find your way out if you get a bit bushed.

Remember, it's not only the mugs that get lost. And sometimes a good man will get into a fix through his cobbers. Most of all we've got to help the children—more and more of them are going out hunting and tramping as the population gets bigger—we've got to know how to put them on the right track.

The bush is one of our greatest heritages—but short of chopping down every tree in the country there's always going to be the possibility that from time to time someone's going to get into trouble. All I'm suggesting is that we think about it and, most of all, discuss it with the children. Then perhaps they'll never feel welling up inside them that terrifying cry, "Help! I'm lost!"

LAKE MAGIC

Tony Nolan

SOMETIMES I WONDER whether we Kiwis really appreciate the abundance of fresh water we have in our open country. I don't mean the sort that comes down in bucketfuls just as you're going to get the cows, or setting off for work on your bike; I mean the rivers and streams and, in particular, the many lovely lakes.

Think of the emptiness of the Australian desert and the Canadian prairie—mile upon mile, with not a hill, not a tree, not a trickle of water as far as the eye can see. And then there's Mexico and Morocco and a dozen other places where the water, if any, is so full of bugs that they ride around on each other's backs.

In a South American country very much like out own, just north of Bogota, there's a little lake formed for a hydro-electric scheme. I sat and looked at that lake for an hour, just as I've sat and looked at lakes at home; as, I suppose, we've all sat and looked at lakes. I realised just how much water means to us here, not only for drinking and growing things, but as scenery: food for the soul, you might call it.

Personally, I'm not so keen on big lakes. A big lake seems to lack a bit of the old magic. Taupo, for instance, never attracts me unless I can cut it down to size, get just a part of it framed by a few trees. I like small lakes; best of all when they're hidden away in the bush and the mountains.

Lakes have something in common with dreams, and there's one lake that will always be like a dream to me because the only time I ever saw it was at night and under a glorious moon. This lake was Rotoma, up in the Rotorua area. I was pushing a bike loaded with gear up the hill from Whakatane on the coast. It was pretty dark and I was dead tired, but I remember the glow-worms in the hollows of the banks beside the road, and how in my solitude and drowsiness they looked like fairy cities.

Then I slogged over the last rise, coasted down the other side and there, right beside me through the trees was a sheet of water as black as ink and smooth as velvet.

I got out my sleeping-bag and lay down on a bit of a beach with my head on a saddlebag and the old bike having a well earned rest beside me. During the night I woke up, to find a full moon in the sky and the whole lake turned into a pool of shimmering silver. As the ripples ran up the beach towards me they sparkled and gleamed so much that I reached out and tried to catch the silver in my hand. It was all so beautiful and dreamlike that I didn't want to sleep again. I left as dawn was breaking and I haven't been back since. In a way I don't want to, because I have a feeling that Rotoma could never be the same in the ordinary light of day.

Another lake that still seems like a dream to me is Lake Ellery, down Haast way. Maybe you can get to it easily now, but in those days it meant a fairly long struggle through thick bush. The Coasters had told me it was a great place for birds—the best morning chorus in the world—so I made it just at dusk and camped on a gravelly piece right beside the water.

Just as dawn was breaking I found myself wide awake, listening. There wasn't a sound—I've never heard it so quiet. But somehow I had a feeling something terrific was going to happen to me. Then out across the water came the glorious notes of a bellbird. It was as if someone had thrown a stone into the sky and ripples of sound were spreading out, like invisible waves from a silver bell. It came again, the call of only this one bird. Then one by one other bellbirds joined in, sometimes altogether, sometimes clashing or a half-beat behind, just as you hear the bells in those big cathedrals overseas. Then in came the tuis and the other birds, until the music echoed back from the bushclad hills across the water and seemed to fill the whole land. Then one by one the birds dropped out of the chorus and it grew quiet again. The sun struck the tops of the hills and I closed my eyes and settled down for a final few minutes' sleep. When I awoke again—the sandflies saw to that—I had a hard job convincing myself that it had all been real.

Another lake that for me has a touch of dreamland is Lake Marion. Marion is one of those coy little lakes that hide their beauty in the forest and reveal it only to the true believers who go to the trouble of looking for it. I'll always remember it because I found it in a rather unusual way.

I knew the lake was near the top of Kiwi Saddle, on the track that leads from the Hope River near Lewis Pass, to Lake Sumner. But I also knew that it was easy to walk past without finding it.

I spent the night at one of Dick Morris's cabins (he was a famous hunter of the Lewis Pass area) and plodded up the track next morning. The track runs through open grass and as I neared the Saddle I saw ahead of me a large patch of countryside of a strange black and white, mottled appearance. I rubbed my eyes and looked again and this time it seemed to me that this patch of grassland was moving, flowing in one piece towards the nearby beech forest.

I stopped in my tracks with my eyes popping out. And then it clicked. The migrating landscape was a huge mob of moulting Paradise ducks—hundreds of them. Unable to fly, they were heading for the safety of the lake. My pack hit the deck with a thump and I was off after them.

I've never seen anything like it. When the ducks spotted me, hot on their trail, they started to squawk like mad and they shot through the trees as fast as their legs could carry them. On the dry beech leaves their feet pattered like rain on an ashphalt road and feathers flew in all directions.

I reached the lake just as the leaders plunged in. All around me streamed the ducks, waddling like a couple of hundred Charlie Chaplins. They didn't seem to realise that what they were running away from was now in front of them, and some of them actually went over my boots and between my feet.

They hurled themselves or tumbled into the water with a splash and swam flat out towards the centre of the lake. Once they were safe they set up such a racket of scolding and swearing that I wished I'd had a rifle to let off just one shot into the air. I bet they wouldn't have been so brave then. But what a lake it was. Emerald green, bush-girt and lovely: that's Lake Marion. And all the more satisfying because of the strange way I'd found it.

It's a great thrill to come upon these secluded lakes in the cubby-holes of our mountains. Some of them are fairly easy to reach—like Lake Daniells and beautiful Lake Christabel off the Lewis Pass road. To get to others you have to be a hard-case tramper or climber. Two rather unusual ones are Lake Castellia at the head of the Wilkin, that fabulous valley in South-West Otago, and Lake Eileen at the top of the Sabine River, in Nelson. Both are cradled among steep snow clad walls and often have icebergs floating on them even in summer.

To go from cold to calorific: the crater lake on Mount Ruapehu can be quite a sight on a fine day in winter, steaming away in a dazzling white bowl of snow under a blue sky. Years ago we used to have a dip in the shallow water at the side and it was a favourite trick to get new chums to coat themselves with a certain brand of suncream to protect their nice pink skins from the volcanic water. When the acid waters got to work on this suncream it turned a dirty black and it just about took a wire brush to clean it off.

Deep down, the water's almost pure sulphuric acid. A while ago some scientific blokes let down a testing device on the end of a steel cable, and all they got back was the top end of the cable.

But my weirdest memory of Crater Lake comes from sitting beside it one sunny day after the Tangiwai disaster, listening to a couple of government investigators talking about ways to prevent the water building up again. One idea, which they discussed quite seriously, was to get the Air Force to bomb the side out of the crater. I couldn't help thinking that the local skiers, already coping with blizzards and eruptions, would find bombing just about the last straw.

Some lakes have been around since the year dot. Some were gouged out by the Ice Age, and others, like Crater Lake and the Tama Lakes, filled up volcanic craters, or places where the earth caved in after eruptions from underneath, like Lake Taupo. But down Nelson way there's a string of them that didn't come into existence until 1929. The Murchison Earthquake of that year brought down whole mountainsides, damming the rivers and ponding up quite large stretches of water. The Karamea, the Mohikinui, and the Matiri Rivers are the best places to see these, and the easiest way is to fly over them on the trip to Westport.

Some of the toughest going I've ever struck is around the sides of those earthquake lakes. The banks are steep, often dropping straight down into the water and, where the slides have wiped off the bush, it's been replaced by a mess of secondary growth, bush lawyer, and stinging nettle. I've seen a strong man sit down and weep—and at the time I felt like joining him.

They reckon that those lakes are full of monstrous landlocked trout. I haven't seen the trout—except a photo of one that I still don't believe—but I've seen some of the smaller eels. We set up camp one evening by the side of a Matiri lake and just for the heck of it some of the boys swam over to the other side. They went like swans but they came back like rockets. When they scrambled out they grabbed an ice-axe each and hooked out a few of the eels that had followed them right to the edge. There's nothing spoils a swim like a nice friendly eel.

Still, I've got off the track a bit. The pleasure of just sitting, looking at lakes. Some of my favourites are Waikaremoana, Rotoiti and Rotoroa down Nelson way, Hawea, Wanaka, Wakatipu. . . .

But the more I think of it the longer the list grows: well known lakes and un-named lakes; bush-girt lakes, and tiny mountain tarns that see nothing from one year to the next but the clouds drifting by above; tourist lakes and little reedy lakes. Lakes where the deer come down to drink and the trout rise at dusk. Kindly lakes and happy lakes, and others, like Lake Browning, high on a mountain pass, lonely, desolate and mist-shrouded, almost like a dream of a prehistoric world.

HOW TO GET GOLD FEVER

Tony Nolan

Gold, gold, gold, gold,
Bright and yellow, hard and cold,
Molten, graven, hammered and rolled,
Heavy to get and light to hold.

When I read that verse by Thomas Hood the other day, I couldn't help thinking how strange it was that the human race should get all excited over a few lumps of yellow metal.

Whole nations have gone wild about it. Countless thousands of otherwise stable people have tossed away everything they valued to go chasing after it into some of the most uncomfortable places in the world.

It's not just lust for riches. There's something about gold itself . . . and digging it out. Fresh from the earth it glows golden, like sunshine, and it has a sort of timeless look. You think, almost reverently, that whatever else changes, in a thousand million years this little lump I hold in my hand will still be gold, waiting to be dug up again out of the earth.

I've seen blokes in the bush wake up at night, get out their little bottle of gold, and sit staring at it. Not gloating, just staring. I've done it myself.

It's funny, but I can't think of anything else that can excite such a wide range of people. Try it for yourself one day. Walk into a room full of ordinary citizens, and casually mention that you struck a bit of gold during your holidays. Hand it round—then watch the gold fever set in.

Last summer I went on a tour of the old goldfields starting up Auckland way and finishing down at Bluff. I'd done a bit of prospecting before, and I'd often wondered just how much gold there really was lying around. I decided that there was only one way to find out—carry out a standard test in as many rivers as possible. So that's what I did.

I had a dig in nearly forty rivers, some around Coromandel but mainly down south, and I got gold out of twenty-five of them: not more than ten to twenty flakes in each case. That's less than a pennyweight and there's twenty pennyweight to an ounce, which is worth up to about $40. So I didn't make a fortune. But I wasn't trying to. I limited the test at each river to half an hour, and I never went more than an hour's walk from a good road. Also, to make the tests comparative, I had to keep to the same sort of place in each river, and that meant sometimes getting nothing, when in a different place I'd have probably struck it. However, it seemed to me to show that for Mum, Dad, and the kids, gold-digging is a good lurk for the next holidays.

What surprised me most during all this galloping around was that gold-digging seems to be a dying art. People are losing the skill. Many I talked to thought it was too technical, or that special gear was needed but, as you'll see, this isn't so.

Here we are at the river. I've picked on one that I know has already produced gold. That'll give me a chance to get my hand in.

For gear I've got a big tablespoon, a shovel—I grabbed the little one by the kitchen fire—and a round aluminium piedish, about ten inches across, with sloping sides and a grooved flange around it. A proper gold dish would be better, or a frying pan, or even the old tin washbowl, but I want to keep it simple. And I've brought a bottle: two, in fact, but the one that's not in the creek to cool is an aspirin bottle to put the gold in. Some optimistic blokes I know carry a jamjar; they reckon it's easier to pop the nuggets in.

So down to the water. There must be some gold here; the place is black with sandflies!

I've purposely chosen a spot just below a fast, narrow bit. There's a reason for that, though. An old timer told me that theory gets you nowhere, because the gold never takes any notice of theories. He's dead right, of course. But like most people I feel happier if I at least think I know what I'm doing.

Theory says that gold, being much heavier for its size than most of the other stuff in the river, takes a fair flow of water to shift. In the big fast rivers it may keep trickling along the bottom of the main current, but in most rivers and creeks I don't think the gold moves until there's a flood. As soon as the volume drops again, or as soon as the gold finds itself in a stretch of slack water, it settles.

Probably most of the heavy gold is right on the bottom of the bed, or maybe on the rock ledges in the big pools. In small creeks you might be able to get at it, but in the big ones it's out of reach. However, we're saved by the fact that during a flood the beaches at the side become part of the bottom, particularly if the water running over the beach is about the same depth as the main flow. Some of the gold may get stranded there, and that's the sort I'm going to look for.

Now the problem is to imagine the river as it was during a flood. That's not easy, and it's harder still to work out where the slack stretches would have been. Rivers can change completely during floods. Of course, you could close your eyes and just dig anywhere and you could be lucky, but then again you might shift a mountain of gravel without finding a speck. I prefer to try to work it out.

So I choose the flatter stretch just below a gorge or a rapid because in most cases that's where the water would spread out and slacken speed. Then I look for a beach that the water would have spread over, or one where the current would have been swung ashore. Then I find a big permanent rock, or a group of rocks that would have further slowed down the water. That's where I reckon the gold will be: on the down-river side of the big rocks or scattered among groups of rocks.

I pick a likely place: say the down-river side of a big rock next to the water, then I beat off the sandflies and start to dig. I don't go very deep, six inches or a foot is good enough. I'm looking for gold from the last flood, not the stuff that arrived a thousand years ago; that's probably up on the terraces above me. If I can't get the shovel in amongst the rocks I use the tablespoon. And I try not to disturb things too much because the gold only sinks further down and you have to dig like mad to catch up with it.

I throw the whole lot into the pan, stones and all. Flakes of gold often hide under a big stone and if you throw away the stone the gold goes with it. With the pan full, I run flat out along the beach—that's not part of the theory, it's to get away from the sandflies—I find a place where the water is about a foot deep and not too swift, and I'm ready to start panning.

I dump the dish on the bottom, under the water, and swirl the dirt around with my hand to wash away the worst of the mud. Then I pick up the dish with a hand on each side, lift it off the bottom—still keeping it under the water—tilt it slightly away from me, and start moving it smoothly in semicircles.

Each semicircle starts close to my body, then goes out to the right, and around. A left-handed bloke might do it to the left. When it's out opposite me, I don't go round the full circle but draw the whole dish straight back into me again. From time to time, I sit the dish flat and shake it from side to side. When I reckon the big stones have been thoroughly washed I dump the dish on the bottom and pick them out.

Then on again with the rotating, tilting the dish a bit more all the time and watching that I'm not so violent that everything flies out at once.

After a while my hands are freezing, my back's aching, and the sandflies are upsetting me more than ever. But the dirt's almost gone from the dish and now I have to be careful. I might even move to a pool, so that a quick surge of the current doesn't whip everything out of the dish . . . I tilt the dish more, but slow down, getting very gentle.

When I can see only black sand in the dish I lift it up, pour off all but a trickle of water, and give it a few little twists. And if I've got any gold it'll be there at the tail of the black sand. Whacko! There it is.

If I'd been too violent, or the current was too swift in the pan, I might have lost the black sand and wasted my time. Or there mightn't have been any black sand in the first place. If I didn't get any I try other spots until I do get it. If I finish up with black sand, even if I don't get any gold, at least I know I'm making a reasonable job of the panning.

There's black sand in nearly every gold river, and usually the gold is with that sand. So I sometimes look for the sand first, either on the ground or in the pan. However, I struck one creek in Otago where the sand was so fine that you could hardly see it, but the gold was there all right, and there might be other creeks like that.

To get the gold out of the pan, press your fingertip on the flake and the gold will stick to it; then put your finger to the top of a bottle brimfull of water, and as soon as the gold touches the water it will dive into the bottle.

Now, in some really big rivers you mightn't find gold anywhere near the water. The volume of water is too much to let the gold settle, so you have to look somewhere else. In at least a couple of big ones the good gold is way up high, on the top of hump backed gravel ridges and mounds of stones, up to thirty vertical feet above the water, because that's the flood level where the current slackens. You can often get flakes though—enough to take home—close to the water but right alongside rough rapids. Thin flat pieces will float on the skin of the water, particularly when greasy or sticking to a disc of black sand—I think it must be an electrical business—and the surges from the rapids toss them ashore. If you see waves lapping up on the bank, have a dig on the uphill side of any rocks that might have trapped the gold.

The gold I've talked about so far is the easiest to get. But of course, you can dig potholes up on the terraces, or borrow a diving suit and fish around on the bottoms of the pools. A bloke was doing well at that down on the Wakamarina, with his wife working the air-pump. A brave man to risk his life like that! Or you can scratch around in cracks in the rocks, either in the water or at the side.

Gold's been found in some funny places: right up on hill tops; in creeks no bigger that the drip off a dishcloth; under the grass roots; even in amongst gorse bushes.

As the old timers say: "Where it is, there it is."

All this gold is alluvial gold. But don't forget the other sort, reef gold. Payable gold in reefs can be specks so small that you can hardly see them. There's plenty of scope for the new-chum prospector in that line. If you find a piece of quartz with a yellow speck or two, follow it up to see where it comes from.

At the start I suggested sticking to the known gold rivers. But it wouldn't do any harm to get in a bit of practice at the creek up in the back paddock or out at the bach. You never know your luck. A lot of gold's been found by mugs who didn't know it wasn't supposed to be there.

Remember what they dug out in the old days? From the Wakamarina, 25,000 ounces. Ten tons in a year from Otago. Twenty tons in a year from Westland. I bet the country could do with a bit of that now to help balance the budget. Come to think of it, so could you and I. . . . Well, what's wrong with a touch of the gold fever?

Camp Ovens — With Recipes Tried and True

TRICKS OF THE TRADE

Linda Gillbanks

FIRST, there are no special recipes for cooking in a camp oven. Anything that can be cooked in any type of oven can be cooked that way: so everyday recipes in any cookery book can be tried. If a camp oven is used over an open fire, it mustn't be the kind of fire that burns things to cinders.

A farmer's wife, writing about the experiences of her family on a sheep station in the Mackenzie Country, said that she baked all the small things such as cakes and biscuits in her household oven, because they couldn't be baked in the camp ovens at the men's quarters. That certainly hasn't been my experience.

The nicest biscuits I've ever tasted were baked in a camp oven: thin, crisp, really delicious oatmeal biscuits, made with dripping, not butter. They were baked by a lone Pakeha woman who lived on a 14,000 acre dry-stock run about midway between Kawhia and Awakino. She was the only Pakeha woman there until I arrived as a bride of nineteen.

My husband was head stockman on that isolated run, and on the first Sunday after my arrival we were invited to have tea at the home of this lady, who had baked the biscuits for our refreshment. Nearby stood another cottage owned by some very fine Maori people, an elderly man and his wife and daughter. The younger Maori woman taught me the finer points of camp oven cooking, and those biscuits were a challenge!

It's nice to have the right kind of wood for the fire, but any old fire will do. If cooking small things such as bacon, eggs, chops, steak, you can go ahead with whatever fire is available. But for slower cooking, such as joints of meat, bread, large cakes and things like that, it's a bit different. In that case it's best to build a good fire and let it burn well down until you have lots of embers. You need ashes too, so don't be over-zealous about keeping your fireplace cleared and tidy.

In the old days we used to bank our fires at night by covering the short ends of wood and the embers with a heavy layer of ashes, *really* banking, with a thick covering around the base as well as over the top.

I believe the idea is to keep out any draught, and if well done the fire will stay almost as it is, and by pushing aside the ashes in the morning the cook has a good bed of really glowing embers all ready to start cooking breakfast. The ashes are hot too, and that's a big help. The hot ashes alone give sufficient heat on the oven lid for slow cooking.

The lid of a camp oven is shaped for holding embers so that there is heat above the contents of the oven as well as underneath. If it isn't convenient to hang the oven over the fire—say for instance that you haven't a good supply of embers, not enough to see the job done—you can stand the oven nearby, leaving you free to build a good fire, and if you use some thinner pieces of wood which will quickly burn to embers, you can feed these on to and around, or under the oven as they become available.

Often when my oven was a bit slower than I wanted and there were no more embers ready, I would arrange what I had to best advantage and then throw on some wood chips and have a little fire on top of the oven. When the chips burned away a very little, on went the ashes and the heat didn't escape.

There are tricks in all trades we are told, and away out back with very little available to get you out of trouble spots, you learn fast.

While the fire is burning down the lid is laid on top to heat. It gets to a red heat very soon, but as it's not needed as hot as that, it can be tested by scraping the end of a stick across it. Though it is still black in appearance it shows a dull reddish colour as the stick scratches it and there's a little trail of sparks. That's about right for cooking scones, but for bread and joints it's usually not pre-heated. The lid is usually lifted by the cook's special "oven stick" to save you from the heat—a stick about four feet long, a sort of suit-yourself length, really. But do watch that balance: ashes and charcoal slide remarkably fast from the lid, and they don't improve any food.

To make things more precarious, women wore ankle-length dresses in those days, and when leaning in towards a fire to hang on pots, kettles and ovens, there was a real danger of dresses or aprons catching alight.

Camp ovens are made in several different sizes, and for bread-baking which took me about an hour and a quarter, I used a large one. They have three small legs about three inches high so they can stand on these legs with space to put embers or hot ash underneath.

When we wanted long-lasting embers, we burned green tawa wood, which burns to thick embers with very little ash. Rata and tea-tree are well thought of. For good heat-giving ashes we burned the hard shell of the black punga-fern. The black punga burns down to needles of filmy embers bedded in ashes. The white punga was of little use—it made masses of ash, more messy than useful, and didn't have the heat that

the black punga gave. When baking bread in my big oven the ashes from the black punga gave all the heat I needed to put under the oven, while around it, a few inches away, I put a ring of embers covered with the ashes and occasionally I'd add more hot ashes. In the hollow round the lid I would put a light row of embers and also part way up over the lid, then I would cover these embers and the upper part of the lid with the hot ashes.

The crux of the matter is regulating the heat, just as we regulate the heat of our electric or gas ovens today. I soon found it easy to judge the heat of my firing, learning a lot from the younger Maori woman in the short space of time we spent on each visit there.

Their living room was a large square room and a small fire burned continually in the centre. Above it the roof sloped up from all sides to an opening and a short chimney. The draught must have been perfect because I never saw that room smoky, although we often made visits.

I used to watch as the daughter tended her camp ovens where they stood near to each other on a band of ashes that almost surrounded the fire. The fire itself spread to a width of about three feet, just a deep smouldering bed of embers and ashes. The woman went occasionally to one or other of the ovens, adding a small shovel of hot ashes on top of or around it as she thought necessary. It was all done so casually while chatting to visitors, either there beside the fire, or outside if the weather called us out.

Remember: anything that can be cooked in an oven can be cooked in a camp oven, and there's nothing to camp oven cooking except regulating the heat.

RECIPES TRIED AND TRUE

Rose Hughes, Remuera; Ruth Gedye, Springvale; Lucy Mills, Gisborne; Nita Schramm, Hokitika; E. L. Brooks, Glen Eden

---

REMINDER: the camp oven cooks best on a bed of embers, with a judicious layer of embers and ashes on the lid (careful when removing lid with a long stick, so that ashes don't slide into the food).

You soon learn to regulate the heat top and bottom, and avoid roaring fires—they're too fierce. Any recipes can be cooked in a camp oven—and *never* put cold water in a hot camp oven: it's made of cast iron, and this will crack it. If downhearted, remember some pioneer women never mastered the camp oven!

JOHNNY CAKES

Johnny Cakes are just plain scones made from: 1 cup of self-raising flour, $\frac{1}{4}$ teaspoon of salt, and enough warm water to make a workable dough. Mix well, and roll out to $\frac{1}{2}$ inch thickness, cut into quarters. Put a few tablespoons of fat in a warm camp oven, and when it is boiling, drop in the cakes and cook 8 to 10 minutes. Eaten with butter and golden syrup, these are something to remember.

ARAHURA PUDDING

I worked this one out when we were miles up the Arahura River and rations were getting low, it went well with golden syrup. 1 cup shredded suet, 2 cups raisins or sultanas, $\frac{1}{2}$ cup of sugar, 2 cups self-raising flour, $\frac{1}{2}$ teaspoon salt.
Method: Mix all together with a large cup of cold tea (no milk). Boil in a cloth (say a flourbag, or in a billy inside a larger one) for about $2\frac{1}{2}$ hours.

Now let's have a real binder, a really good:

CAMP OVEN STEW

Cut up the required amount of meat (I use deer or wild pork), brown this in the hot camp oven with a little fat. Then add enough warm water to cover meat; add plenty of onions (whole); carrots cut up; and 1 cup of dried peas which have been soaked overnight. Salt and pepper to taste.

About half an hour before this is cooked, I make what my family calls "B-------s Afloat", but we'd better call them:

SINKERS

1 mug flour, 1 teaspoon baking powder, ½ teaspoon salt; mix together, then rub in 1 tablespoon butter. Mix with milk and water to a soft dough, make into small balls and drop into boiling stew. Cook half an hour with lid off. If need to thicken, use a little oatmeal.

VENISON

Grease your camp oven fairly liberally, and set it over a quick fire to get good and hot; and on that sizzling hot bottom of the camp oven sear your slices of venison, about ¾ inch thick, quickly but thoroughly on both sides, browning them nicely.

Then, over your seared steaks, sprinkle one or two handfuls (depending on the size of your fist and how big a brew you are making) of rolled oats mixed with about a teaspoonful of dried herbs; next a layer of sliced potatoes, then a layer of sliced onions; then a tin of peas, liquor and all, then a tin of tomatoes, again liquor and all (or three or four fresh tomatoes).

Season well with salt and pepper as you go, and last of all, if necessary, add water so that the liquid just shows beneath the top layer of vegetables.

Now on with the lid and swing the camp oven over a fire that will keep the contents gently simmering for an hour or so, by which time all should be cooked in a thick brown, nutty gravy.

P.S. Just a note to young shooters. Please take pains to bring your venison to the kitchen in reasonable condition—not covered with sticks, stones, hair and gore and other unnecessary and unsavoury ornaments. Honestly! Considering the horrible state of much of the hard-won booty that comes home, it's no wonder so many folk—especially the women folk—wrinkle their noses and say, "Ugh, *venison*!" Please take the trouble to wipe over or trim off any doubtful-looking surfaces and edges before presenting that haunch or leg or steak or whatever to the cook, and then you and your offerings will be greeted with enthusiasm on your return from each trip into the bush.

POTATO SCONES

Mash boiled potatoes till quite smooth, adding a little salt. Sprinkle board liberally with flour and knead mashed potato into it several times. Roll out to thickness required—about ¼ inch—cut into oblong pieces—prick well with a fork after placing on hot greased camp oven. Do not cover. Turn with knife when browned on one side. Delicious spread with butter and eaten hot.

ANY SCONE MIXTURE

Place scones, whole or cut, on bottom of hot greased camp oven. Cover with lid. Good heat under camp oven, plenty of glowing embers on top. Turn oven round occasionally to ensure even baking.

DAMPER or BROWNIE

For Damper: 1 lb flour, 2 teaspoons bicarbonate of soda, 4 teaspoons cream of tartar.

Rub in with fingers, 1 tablespoon butter. Mix lightly with 2 breakfast cups milk, or water.

Make into round loaf, place in greased camp oven over good heat for $\frac{3}{4}$ hour or more—lid on.

For Brownie: add $1\frac{1}{2}$ cups sugar, 1 cup sultanas (or other fruit), 2 tablespoons golden syrup, 2 eggs, and 1 cup butter or dripping to the Damper dry ingredients, before mixing with the fluid. When eggs are used, less rising is required. Lid on.

PUFFDALOONAS

Your camp oven will need deep, well boiling fat (no water). Drop in pieces of any scone mixture and cook till crisp and brown. Do not cover. Eat at once, with meat dishes and gravy, or alone, with butter or honey.

ROTI OR OATCAKES

An Indian Recipe

Knead the oat-flour with water and roll out in round cakes as thin as the edge of a penny, and 6 inches in diameter. Bake slowly on a greased and lightly floured camp oven. Turn with knife or spatula—no lid. A little melted butter is brushed over before serving. (Roti is eaten with meat and vegetable curries in place of rice in some parts of India.)

POTATO BREAD

When straining potatoes, keep the water. When lukewarm, add $\frac{1}{2}$ teaspoon compressed yeast, $\frac{1}{2}$ cup sugar. When well mixed, add 1 cup flour. Stir thoroughly. Let this rise in jar (with light covering) until morning. Then mix 6 cups flour, 1 pint warm water, $1\frac{1}{2}$ pints potato yeast (from jar) to form a "sponge". Stand 3 hours in warm place. Work till it leaves sides of bowl. Then leave 3 hours more, working it down twice. Place in camp oven, and when well risen again, cook slowly for a soft sweet crust.

MR BROOKS'S QUICKIE

While the camp oven is being prepared for the main meal, put your companion and that billy to work for a quick second course for two.

Throw ½ cup rice in a billy, with a pinch of salt and 1½ cups water. Put on fire to boil while main meal cooks. Then take billy off fire, toss in ½ cup raisins, sultanas, dates, or mixture, add two desertspoons sweetened condensed milk. Stir thoroughly and add a little more water if needed. Put back on fire to simmer, stirring now and again. Ready in 15 minutes.

Kids

THE LONG PADDOCK

May Hill

WE MUST HAVE BEEN short of feed that year.

As far as I was concerned, this wasn't such a bad thing; I was excused my usual Saturday afternoon work in the farmhouse, and sent instead to look after our dairy herd. Twenty-odd milkers, allowed to graze in "the Long Paddock".

The Long Paddock was our two-mile stretch of country road, gravelled at last in the centre, but with plenty of cattle fodder along either side where the grasses and weeds grew in wild profusion.

The cows seemed to enjoy the change from their usual pastures, those neatly fenced, strongly secured paddocks where it was impossible for even the most venturesome and inquisitive member of the herd to stray, however much she may have longed for such freedom.

So no wonder the cows, when turned out on to the road, galloped at first in highlighted abandon, their hooves raising clouds of dust as they hurried along, in a way rather like a crowd of children released from the school gates and free from adult supervision.

Such behaviour could possibly be excused on the part of the young skittish first-year calvers, not yet settled down to sober ways and the serious business of keeping up appearances as well as the milk supply, but most unbecoming to the usually serious-minded, more sedate matrons of the herd, dashing along with undignified haste.

Realising that it meant a welcome change from their usual paddocks, I felt a certain sympathy for this boisterous hustle and bustle. But it meant some extra work for me; I'd been sternly commanded to head them off before they reached the junction of the dangerous main road.

Often I had to run at top speed, and sometimes the leaders were almost at the end of the gravel and quite prepared to trespass on to the tarseal if not turned back in time. I would be gasping and puffing, and wondering why I was never allowed to use Barney, the capable dog-of-all-work about the place.

I think the grownups had some strange idea that Barney would become overexcited and make the cows get out of control, so they thought it wiser to leave the whole job to me since, as they fondly imagined, I had nothing else to do.

How utterly wrong they were!

I had a thousand more or less interesting things to see and do, in those hours spent herding cows on that semi-deserted stretch of country road.

First I had to check on the fascinating insect world, so enchanting and enthralling that I can find no better way of describing how I felt than the apt quotation from Scripture:

The eye is not satisfied with seeing,
nor the ear filled with hearing.

If you have never experienced the thrill of seeking out with gentle fingers the fragile, delicately-painted ladybirds, their tiny backs lacquered orangey-red with black polka dots, you have certainly missed something.

Diligently I searched along the row of fenceposts, where a kind of grey lichen grew quite thickly into the cracks and crevices in the wood; a favourite place for hunting ladybirds.

Each little insect was put in the palm of my hand, studied carefully, then allowed to walk with a delightfully ticklish sensation across my fingers and back to its refuge in a post or strainer.

This ladybird-hunting was exciting because you never knew just what you would find. Mostly those ladybirds had two black dots, but now and then came the thrill of finding one with no less than seven spots, and once, on a never-to-be-forgotten day, a remarkably handsome ladybird dotted with the grand total of twenty-four neat black dots on her shiny body.

That was something to tell the kids at school on Monday!

Trouble was they'd never believe me, so I half wished I'd kept that wonderful specimen to prove it; but I knew that she was much happier safely tucked away in her private hideyhole.

And then, of course, there were the crickets to be stalked, with utmost care. Scarcely daring to breathe, gently, gently, picking my way cautiously over the rank growth, nearer and nearer, my unsupecting quarry swinging nonchalantly on a swaying dried grass stalk, chirping his cheerful notes, over and over again the same tune.

Ha, got him. A moment of triumph as he flutters in my imprisoning hand. A scared cricket with a neat greenish or red-coloured body and gauzy wings. I had no intention of harming him. The thrill of catching one and allowing him to relax, then walk with his dainty feet right along my bare arm before flexing his wings and flying away, was sufficient for me.

There were butterflies too, but I don't ever remember seeing one of the destructive white ones which these days rise in a fluttering pulsating cloud from a field of turnips or greens, destroying thousands and thousand of plants.

Along the roadside was an abundance of the common black-and-white butterflies, and now and again a gorgeous brown-and-orange beauty, to be hopefully pursued but never caught as it fluttered tantalisingly just out of reach, finally swooping effortlessly away over fences and hedges to disappear from view.

Since the herd usually settled down after a while to grazing peacefully, I had plenty of time for my own delights, only now and again having to restrain old Black Bess, or huge rawboned Hetty the Holstein, from a sudden urge to explore beyond the bounds of safety.

Even in those far-back days there were quite a few cars on the main highway, some of them driven at the reckless rate of thirty or forty miles an hour. Their drivers could not reasonably be expected to stop suddenly or swerve to avoid the large bulk of a matronly dairy cow wandering on the main road.

How like people these creatures were.

Idly gazing over my charges, one by one I mentally compared them with people I knew. Bella, the leader, for instance; always bossy and domineering, expecting others to follow where she led. Why, there was a girl in my class exactly like her. And Queenie, the sleek light-coloured Jersey, pride of the herd and well aware of it, reminded me of a certain spoilt lass in the sixth standard. Oh yes, they were like people all right; the inquisitive, nosy ones, always fossicking about; the shy, quiet, inoffensive types; and the downright lazy ones, glad to eat and then lie down to sleep.

Some were timid, easily scared, and would kick the milk bucket flying if startled, and some were the unimaginative, impassive ones who didn't mind in the least what was done to them.

I came to know the cows as individuals during the hours I was their custodian, and some I grew very fond of; but, just between ourselves, I never could bring myself to the point of liking Old Whitey with the horrible hairless sunburnt patches shining pinkly on her rump. It wasn't her fault she was so susceptible to sunburn, but I never liked even to touch her.

Here and there by the barbed-wire fence of the roadside stood patches of bush, and sometimes I'd scramble over and climb up into the shade of the leafy branches, investigating the birds' nests, empty by this time of the year except for an occasional rotten egg. The young thrushes and sparrows had long since stretched their wings and flown out into the world.

In a swampy place a clump of toetoe thrived with its long straight stalks and feathery flaxen heads, ideal for picking a handsome charger to mount and then to gallop along, shouting and yelling at the cows in the approved Wild West manner. Mind you, my glimpses of Wild West doings were stolen from the books devoured eagerly by my grown-up brothers and strictly forbidden to me, a girl, but I knew enough to imagine myself a lady sheriff bringing in the outlaws with a flourish.

Those hours spent with the cows enjoying their free grazing in the Long Paddock went very quickly, and often I was surprised when the whistle of the four o'clock train, signal for me to take my charges home for milking, sounded loud and shrill. The time had come for me to leave the Long Paddock, to leave my magic world.

THE WAY WHEREIN THOU HAST BEEN INSTRUCTED

John McCaw

"LET US ASK a blessing." So said my father before every meal.

There were eight kids in our family, a healthy bunch and always as hungry as a drover's dog. As far as we could see there was only one good point about this business of "saying Grace"—it meant a good clean start off the same mark for everyone.

We had a pretty good idea of how long Grace would be at each meal, because it always depended on one of two factors: the social standing of any visitor, or the quality of the meal.

My father was no social climber, and it wasn't wealth, title or distinction, unless it were in the church, that counted with him. It was his depth of affection. Grace could run to twenty minutes if one of the family was off that day to boarding school, or if an uncle or aunt had come for lunch. If a brother minister were present, ten or fifteen minutes was a sure bet. A decent meal without visitors was good enough for five minutes, but cold meat and fried spuds deserved only two minutes at the most.

There was one splendid occasion when some close friends were present, a specially good meal was spread, and Grace went on and on.

The youngest present, aged about four, left her chair, took the orator's hand and said sweetly: "Gran'pa, little girls get terribly hungry!"

The Old Man had a wonderful sense of humour: he ended his prayer abruptly and collapsed in a gale of laughter.

As the family grew older and timetables for buses or trains had to be kept, morning family prayers were more and more difficult to arrange, so the Grace at meals tended to take their place.

The older generation of Scots Presbyterians, though emancipated from the sternness of their covenanting forebears, still clung to the recitation of the Psalms, the Shorter Catechism, and long Bible readings, all a bit boring to the young. But the old people had a magnificent, childlike faith in their God. He was in everything, not as a sort of Gestapo agent, but as the motive power of every action. He it was who controlled the wonders of the universe, the orbits of the planets, the movements of the galaxies; but He, too, directed the ways of man and beast. He caused the cow to begin its lactation, the vegetables to grow, and even the blights to mar the potatoes.

The *New Zealand Farmer* of 1890 tells the story of a Hastings orchardist who was chided by the Agricultural Advisor of the day for not spraying his trees against the codlin moth.

"Ah," said the orchardist, "the Lord sent the moth, and when it pleases Him He'll take it away."

My old dad was not as silly as that. In a similar situation he would have placed the problem before the Lord; and then he'd have gone out and sprayed or painted with all the vigour of his by no means feeble frame.

He was a massive man, about All Black forward standard. Like Moses of Old Testament times, these old Scotsmen talked with God and, like Moses also, they were not above arguing with Him too. Our breakfast and dinner tables were the setting for these conversations—one-sided, we thought them, but not so the head of the house. God was presiding, and my father would tell Him all about the doings of the day. What had happened and what had yet to come. He thanked God for all that had happened, good or bad, and asked for guidance in the future. The good or happy events were attributed directly to Him, only the unpleasant were our own fault, and confession of this was made.

My dad and his God were great friends. The discussions would range from politics to poultry—with more value placed on the poultry; from sermons to surgery; porridge to public houses. He would mention us all by name, one with his studies, one with a change of job, one on holiday, or one unwell. This was an excruciating experience for a teenager going through the agnostic stage, and called for great long-suffering from hungry youngsters.

But Grace was invariably shortened by a meal served with white sauce or a milk pudding. This was the Achilles' heel: there was nothing to be thankful for in this "clitter" as old dad called it. About ten words in a grudging tone sufficed for "clitter". Later on in my career I went with a client to do business with a very careful old bachelor fellow away out in the sticks. He invited us in to lunch, spread a lavish feed of stale water-biscuits, without benefit of butter, and milkless tea before us; then spent five minutes giving thanks for the food. As my irreverent friend remarked when we left: "I wonder how long he'd have gone on if he'd had anything to be thankful for?"

Family prayers in our home were by no means an unhappy, boring or distasteful time to us—far from it. For though we were always full of go and energy our parents made the morning worship a time of tremendous interest. One, usually dad, would read a passage, usually from the Old Testament and chosen with care for its appeal to children and its moral lesson. What great stories there are in that book, and how much poorer kids are today for the lack of them.

Poignant stories of Joseph and his brothers, blood and thunder with Joshua and Gideon, intrigue and vengeance with the women Jael and Delilah, the great physical strength of Samson and his weak moral fibre, the dramatic end to his career. Elijah taunting the political leaders of the day, or boldly accusing the King of treason; David a fugitive from Saul, and his final triumph; Solomon and Amos; Ruth and Esther—wonderful stories. What fortunate youngsters we were.

How well these stories were read—the fighting and dramatic passages in a thunderous bass, the lilting love stories in tender tones, the intrigues and plottings in eager haste, the horrors accentuated by careful pauses and whispered words—but always with a deep sense of the reverence due to the Word of God. There was no moralising at the end; the lesson had been told in the story by the careful modulation of the voice and the emphasis on the crucial passage.

I remember as a small boy I could scarcely wait sometimes till the next morning to see what happened next. How would Joseph get out of that pit? Or Jepthah get on with his vow to sacrifice the first living creature he met after his victory—and then met his daughter? Or poor young Isaac bound to the altar with his father brandishing a knife over him? What next? Paul left for dead after a stoning, or hanging in a basket on the city wall, or bitten by a deadly snake. How would he survive until the next episode? How we wept with the same Jepthah and laughed at Elisha when he made an axe-head swim! We learned the arts of sheep breeding from the tricks that Jacob played on his father-in-law Laban, and we cheered loudly at his final triumph in securing Rachel for his wife. We marched with Gideon and David, we raced with Jehu, and we taunted the heathen priests with Elijah.

"Call him louder," we yelled. "Perhaps Baal is asleep."

The recitation of the Psalms was not so welcome. The words seemed old hat, and often we did not get the meaning. But the cadences stayed with us and our later years revealed the meaning. We learned to weep with the Jews on the banks of the Euphrates, we laughed with them at Jordan, we exulted over the defeats inflicted on the Hittites, the Jebusites and the Perrizites. As for the Philistines, we loved to vilify them in stirring sentences culled from David's songs.

"Let hot burning coals fall on them, let them be cast into the fire and into the pit, that they never rise up again."

"Amen," we chanted fervently.

The thought of oil anointing Aaron's beard and flowing down to the hem of his garments filled us with amusement.

When we had been found out in some mischief, like raiding an orchard or pinching golf balls from the nearby links, we would quote, "O Lord, chasten me not in Thy hot displeasure."

We swallowed the story of Jonah as easily as the whale swallowed our weak-kneed hero. When Father wasn't in earshot we sang songs about him (and others):

Now Jonah he was sent
To make Nineveh repent. . . .

and finishing up with:

And when the fishy atmosphere grew heavy on his chest,
Old Jonah pressed a button, and the whale—he did the rest.

Our theology was simple and solid. If our God prepared a fish to swallow Jonah—well, that was OK by us. If he wanted to, He could.

The Catechism was not so good.

That meant solid memorising—lots of repetition, parrot-fashion, with long strings of unfamiliar words: efficacy, ordinances, foresignify, covenant, election, reprobate, and suchlike. But learn we did, though I must confess they are pretty well all gone from memory now.

But the prayers at morning devotions were quite good fun, provided you were not itching to be off somewhere in a hurry.

There was always a bit of a scramble for the best chairs—we knelt for prayers—as the high-backed chairs with slats were greatly preferred. You could see through the gaps without being observed by parents or aunts. Uncles didn't matter—they often looked themselves—but aunts were the very devil; they seemed to be trying to find trouble, and with our big family they found plenty of it. According to them we were all on the road to hell, going downhill at slalom speed. I'm sorry to admit that we pictured hell as a particularly unpleasant place, policed by elderly spinster aunts.

Peeping through the chair slats was more unobtrusive than peeking through one's fingers and, provided the chairs were nicely placed, you could make faces at another urchin, poke out your tongue or even, after a careful glance round, put your fingers to your nose. Quite effective when done through the narrow opening in the chairback.

Failing a high-backed chair, a low-cushioned one was the next best. For on this you could sprawl out comfortably while the patriarch roamed the world with his Creator looking things over. Prayer was offered for the King (Edward VII and George V in my day), the Prime Ministers of Great Britain and New Zealand, both Cabinets (was a cabinet a Parliament or a cupboard?), all politicians at home and abroad, the Emperor of Russia, the President of the United States, Royalty in many countries, the local Mayor and Councillors, policemen and railway guards, soldiers and sailors (no airmen then, or they would have been included), doctors, nurses, and brother ministers. Even priests were included, if the feeling was ecumenical at the moment.

The family were named one by one in order of seniority, with special pleas at exam times or in sickness. If one of the family was travelling, he was commended to the care of God, who apparently travelled with him if asked.

There was always cause for thankfulness if and when the outlook was grim and foreboding. I can clearly recall the prayers of thanksgiving at the time of the 1918 influenza epidemic, when three of the family died and two were critically ill with little hope of survival. There was no note of complaint nor hint of hardship, no abysmal depths of sorrow were explored. But my father expressed his gratitude for his own personal physique, so that he could continue with his work of succouring the hundreds of ill and comforting the dying and the surviving mourners.

Happy memories of each lost child were recalled, and thanks given for their lives. Pleas were entered for the healing of his own two sick, always ending with the quotation: "Nevertheless, not my will but Thine be done." And so the conversation with God would go on. The congregation would be prayed for, its problems laid upon the table, the church mission field and the social service societies. No Rotary or Lions then, but the Red Cross, St John's Order, Mission to Lepers, British and Foreign Bible Society, Orphanage and Old Age Work (as it was then known) were all remembered.

And all the time the youngsters' minds ranged away to their own interests, but often were brought back by the very diversity of the projects in prayer, the earnestness of the supplication, and the friendliness of the dialogue. For dialogue it was: there was listening as well as talking.

The discipline irked us, yet we did not rebel. The stringent rules of the Sabbath Day, though now considered an object of mirth, did nothing to mar our enjoyment of life; rather, they gave a keener appreciation of the many liberties we were given. Because we could play no games on Sunday we only enjoyed our Saturday ones the better. Since we could have no reading matter except the Bible or a few special publications, we found other avenues of pleasant relaxation. We sang round the piano; we did crossword or similar puzzles, based on religious or Biblical themes; we engaged in endless friendly argument, far from theological areas; we had dozens of young friends in to meals; we swam, and we rambled the country side. Simple pleasures, but made the more delightful by the ever-at-hand companionship of loving parents and a big family.

I've been whacked for whistling on the Sabbath. But surely that is better than being allowed to desecrate the day by whooping it round the countryside on noisy motorbikes, or boozing on beaches in broken-down bombs.

Family Worship always, and Grace over meals frequently, ended with all present joining in the Lord's Prayer.

"Our Father," my Dad would say, and then to ensure we all joined in at the beginning, he would pause and start again: "Our Father . . ." This always caused my mind, when I was a wee boy, to conjure up a picture of "Our Father—in heaven."

Always He was a big man like my father, bearded like my father, and with a similar deep bass voice with a strong Scottish accent.

And so I still see Him sixty years later.

"The way wherein thou hast been instructed. . . ."

I am thankful for *my* way.

GUMBOOTS

Patricia Clemett

THEY WERE WAITING for us every morning when we opened the door leading from the kitchen on to the back verandah, lined up in a tidy row, ready to start the day's work with us. We could find nothing glamorous about this row of objects, yet every member of the family would grab eagerly for his or her favourites and proceed, not so eagerly, in the direction of the milking-shed.

It was always a case of first up best dressed at our place, especially when it came to gumboots; and who, I ask you, would want to be left with the old leaky ones? Still, on the farm gumboots were a necessity and someone had to strike them, and once at the shed the early birds, with heads against the side of a cow and with milk swishing into buckets, wouldn't bother to glance around to see who the unlucky members were. We knew from past experience that we would hear soon enough, for sooner or later someone would have to step out into the muddy yard to drive a reluctant cow into the bail, and language hot enough to vulcanise the hole in the offending boot would pinpoint him or her.

Visitors calling at the farmhouse and knowing that our ages ranged from ten to twenty years would look at the boots and remark: "How do you know your own boots? They all look the same size to me!"

No wonder. They *were* the same size, all 8s. Dad believed in buying boots for us to grow into, not out of, and the younger members of our family found it quite a problem keeping them on, but we soon learned to walk stiff-legged and with toes curled up. We usually made the grade; that is, unless we had to turn to round up a wandering cow. Then and only then, would we part temporarily with them. On the other hand, big-size boots did have certain advantages. I've heard Dad roar with rage and pain when a stubborn cow planted a hoof fairly and squarely on the toe of his boot, but when the same thing happened to us we could withdraw an unharmed foot, give the cow a hefty shove, retrieve the boot, and carry on with the job of leg-roping the brute.

It seemed our jobs on the farm depended on the boots we were wearing. The two with the leaky boots had the task of feeding calves and pigs. The one wearing a good pair got the job of carrying water from the nearby creek to wash out the floor of the cowshed, and the one wearing the only pair of thigh-length gumboots automatically took the cows across the creek to be shut in a certain paddock for the day.

We had names for our gumboots.

I remember when my brother Laurie, who should have realised by this time that feeding pigs meant leaky boots, unthinkingly stepped into the muddy sty to right an overturned pig trough. The look on his face when he realised his mistake was worth recording on film, but incidents like this always seemed to happen when a camera wasn't handy. These gumboots, although scrubbed and disinfected inside and out, always retained their own particular smell; and so they became "The Stinkers". The thigh-length ones were, of course, "The Waders"; the smooth-worn-soles ones were referred to as "The Slippers"; the cleated-soled ones were simply "The Cleats"; and the short ones were "Stumps". Eventually "The Stinkers" were replaced and put aside as "Spares".

Our dislikes of certain gumboots were strictly seasonal. In the spring, when the whitebait were running, everybody wanted The Waders; and later, when the fruit in an orchard across the river was ripe, they were always popular. And Dad, when he wanted to lift his flounder net from the mouth of the river, would make sure he had The Waders on before he called for a volunteer to go with him.

We were all surprised when Laurie began to wear The Stinkers again. Surprised, too, when the boots became a little bit shorter each day. The reason Laurie gave was that he was just trimming away the rough edges; but one Saturday morning three of his schoolmates turned up at the farm demanding the shanghais he was making for them. Oh, well, I suppose three shanghais for sixty marbles was a pretty good swap.

Our cowshed was quite a distance from the house and stood fairly close to a small river. Usually one member at least of our family would wander home, after the milking was finished, along the riverbank. Imagine our delighted surprise when we all arrived home to find older brother Barney carefully picking bits of twig from a bowl of beautiful whitebait. The first of the season! We had them for tea, and wondered what sort of season it would be. Even at sixpence a pint, with all the family fishing we'd make a quid or two.

Two days later Mother was already home when Barney arrived with about three or four pints. They were promptly fed to the hens, while a protesting Barney exclaimed heatedly, "But you ate them the other day, and I brought those home in my gumboot too!"

At our place gumboots were never, never discarded until they were absolutely useless. One pair of old boots we could always look at with appreciation was a pair that Mother had painted, filled with soil, and hung on the front verandah with any trailing or climbing plants she could lay hands on.

Always that wild rush in the mornings to grab and don our chosen boots; but we learned at last to look *in* the boots before putting them on. That was only after a succession of finds: a half-eaten apple, a soggy wet fisherman's sock, a fishing line complete with hook and bait, and often, especially after a rainy night, one of those horrible bush insects known to us then as taipos and later as wetas; also the things which didn't get into the boots by accident: a carefully set mousetrap, a couple of heads of Scotch thistle, and once a very live hedgehog.

Still, our family owed a lot to the gumboots. Working around the cowshed would have been rather unpleasant without them. They were invaluable in our eager early-morning searches for mushrooms. Plentiful supplies of flounders and whitebait, and rows of jars of blackberry and apple jellies and jams, all gave proof of where we'd been with our gumboots. So why, I wonder, have I developed that certain "thing" against them? Perhaps it's because I've had my problems with them; a recurring dream haunts me, in which I'm running, running to catch a bus—only to find myself weighed down by an outsize pair of boots. Or, a dream where I find myself in the midst of a social gathering, suitably attired—except for my footwear.

And thoughts that go back to farming days remind me of Dad's remarks when one of the girls would mention that she was bringing a certain young man home to meet the family.

His first inquiry wasn't "Who is he?" or "Where did you meet him?" But always, "What size boot does he wear?" And you can bet your life Dad would find out at milking time.

We all went through it, all four of us. Small wonder that of all the romances which began on the farm mine was the only one which had a happy ending. Dad must have liked Bill; he could already milk cows, he wore size 8 boots, and, most amazing thing of all, he wasn't offered The Stinkers to wear.

The gumboot problem still goes on. My eight-year-old daughter insists on wearing them around home in the summer, and goes barefooted in the winter. My husband works in a job where thigh gumboots are often used, especially during the winter, and it falls to my lot to strap those boots together, sling them over the clothesline, and hose the mud from them.

The years come and go, but it seems gumboots will always be with us. The fashions may have changed a little for those people who live nearer town, but out on the farms where they are the farmers' best friend gumboots aren't much different from those of forty years ago. They'll continue to plod their way through mud and slush and a few other things I won't mention, through long wet grass, through rivers, creeks, swamps, blackberry and gorse, up hills and down dales—abused, misused, but still dependable and indispensible to the man, woman or child who sloshes around in our open country.

WHERE HAVE ALL THE BIRDNESTERS GONE?

Bennie Thomson

WE CHILDREN used to make pocket money selling birds' eggs. We lived in a grain-growing area, up the Old Renwick Road, out of Blenheim; acres and acres of wheat, oats, and barley.

We collected spugs' eggs, as we always called them, and sold them to Mr McCallum. I can't remember what eggs other than sparrows we sold, probably blackbirds and thrushes, but not starlings. Sparrows were always called "spugs" or "spadges".

I've been doing a bit of research about all this, talking with Tom, an old Marlborough friend, and with Roy, who lived in Papanui when he was a lad.

Tom tells me that in those days, instead of County Councils there were Road Boards, and the country was divided into ridings. Mr McCallum, to whom we sold our eggs, was Chairman of the Omaka Riding Road Board.

I thought he gave us twopence a dozen for them, but Tom tells me a penny a dozen for eggs and twopence a dozen for heads. There were only girls in our family, and I can't remember ever selling heads, but I could be mixing up my memory of egg prices with the price some of the boys in our area got for heads.

Tom's best customer was Mr Ham, a neighbour in Alabama Road.

"We boys would buy at half price from further away, and make a profit when we sold the eggs to Mr Ham."

Tom confirms that starlings' eggs were not bought—"unless we put a few black dots on the big end and sold them as thrushes' eggs."

When they had some birds' heads for sale, Tom and the other boys liked to keep them until the blowflies had been at them. Then Mrs Ham would not be so sure of the count! She would count them into the big milk-barrel where the pig food was kept—eggs were counted and thrown in here, too—and many a time Mrs Ham paid for some of the same heads and eggs over again.

Roy tells me that when they sold heads these would be thrown into

a tin out the back. Roy and his friends would sneak round, retrieve some heads, and sell them again.

A friend brought up in Marton told me that the Rangitikei County Council bought eggs and heads; he and his friends also used to sneak around to the back of the Council member's house to retrieve and resell eggs and heads!

Mr McCallum used to take the eggs we sold him round the back of his house to a rocky hollow. There he threw them down, good and hard! They were quite irretrievable. Maybe he just knew the tricks of the trade. Perhaps he himself had been up to the same rackets when he was young?

I can still recall the feelings I had as we stood and watched Mr McCallum throw the eggs down into the hollow. I don't think it ever crossed my mind that we could have sold them again; it was just that, when you've paid good money for something—well, it's just vandalism to smash them like that and not do something constructive with them.

Everyone with whom I've talked birds-egging seemed to have much the same methods of getting the eggs down. Sometimes a sparrow's nest could be reached with a long stick, so you just pushed it down out of the tree. This method could not be used with the bowl-shaped nests of thrush and blackbird.

When we had to climb a tree to reach a nest, the safest way to carry the eggs down without breaking them was in our mouths. We all did it. Now we're grown up, we're horrified at something so insanitary, but then we thought nothing of it. And sometimes you'd be lucky enough to have eggs given you by grownups who were cleaning birds' nests out of downpipes or guttering round the house.

Tom and his friends had another shrewd idea. Many birds lay four, five, or six eggs, and then sit on them. These boys discovered that, if they took one egg only each day, the bird might be persuaded to lay as many as a dozen eggs before she lost heart.

How did the boys catch the birds for their heads?

Roy tells me that on their property at Papanui they had big poultry runs with wirenetting fences. On the top rail of the fence flocks of birds would come and perch when the hens were fed. So the boys would rig up a pea-rifle at one end of the fence rail, fasten it firmly, load it, tie a long string to the trigger, and keep out of sight with the other end of the string in hand. When there were plenty of birds sitting along the rail, they'd pull the string.

Tom remembers that grain crops were stacked and later threshed. The straw was kept to feed out to stock in the winter. Birds would camp in the straw at night—and many of them lost their heads.

Sometimes a wooden frame with netting was propped up on a stick, and some grain scattered under it as bait. A long piece of string was tied to the stick, and with the other end of the string in hand a boy would hide until the birds came for the corn.

One good session with a trap of this kind might be worth a hundred spugs' heads.

Roy used a trap like this, but with the springy thin willow twig on which the birds would perch and collapse the frame on top of themselves. There was also a similar way of doing it with several bricks.

Tom mentions that instead of taking the eggs from a known nest they'd sometimes leave them to hatch—heads brought twice as much as eggs.

He remembers that the County would vote about £15 a year for poisoned wheat, which would be given out to farmers by the local Councillor in the winter, when food was short for the birds.

These birds, originally brought out to New Zealand for sentimental reasons, became such pests that in 1882 Parliament passed the Small Birds Nuisance Act—"to authorise local governing bodies to levy rates for the destruction of sparrows and other birds injurious to crops". This Act was repealed in 1889 and replaced by a new Act, itself repealed and re-enacted in 1902 as the Birds Nuisance Act. In 1908 it became the Injurious Birds Act. That would be the one we worked under. It was finally repealed in 1953. It said, amongst other things, that "poisoned grain may be laid". Another interesting thing is that one of these Acts gave special powers to the Hawke's Bay County Council to deal with rooks.

Someone in Blenheim, who was on the Council in the days when all this was going on, mentions the price for a kea's head was 7s. 6d. Keas, in the high country, attacked the sheep. He also tells me that the average number of birds' eggs bought by the Council in Blenheim over a period of nine years was over two thousand dozen a year. No wonder it was a good source of income for us.

Not all of the children in Marlborough were so lucky. My friends who lived in the Picton Valley had no market for birds' eggs and heads. It was only in the grain-growing areas.

The chief grain-growing provinces were Hawke's Bay, Wellington, Nelson, Marlborough, Canterbury, Otago, and Southland. So there must have been other parts of New Zealand besides Malborough, Canterbury, and Rangitikei, where birds had a price on their heads—and on their eggs.

This type of pest control may sound barbarous. Nowadays we are much more refined—DDT and all that!

OC 10

MACKENZIE WINTERS

Christina MacDonald

A LITTLE WHILE ago I went south to the Mackenzie Country where I grew up. As I drove along, the familiar scenes brought back the frosty ghosts of the bitter cold winters I had known.

The air so cold it made you gasp as it caught your breath; it went straight through the warmest clothing. Feet without any feeling, stamp as hard as you could; fingers so numb you could hardly unsaddle your pony. We grew as surefooted as cats, keeping upright on icy paths: and patches of some paths never saw the sun all winter. We were told that in Victorian times children in Scotland went to school clutching a hot potato which had been baked in its jacket. It kept their hands warm enough to write, and provided their lunch. I often wanted to try out the first half of this, in those Mackenzie Country winters.

Frosts each night were between 10 degrees and 32 degrees—or zero —I have known frosts to be below zero—often it was only above freezing point between 10 am and 2 pm. Then we could feel the warmth of the sun, but only when out of the wind, that bleak, bitter wind sweeping down from the mountains around our home. It killed all the warmth, stung at your face, caught your breath, and numbed your hands and feet in minutes.

We had to break the ice many times for stock to be able to drink, and after a snowfall we went around the garden with walking sticks and banged masses of snow off the branches of smaller trees and shrubs, or the weight would have broken them off.

In the early winter the dog kennels were packed full of straw, and it was fun to watch the sheepdogs go in head first, then turn round and round until they had found a deep hollow place at the back, an atavistic return to the ways of their wolf ancestry.

I remembered darning some stockings and leaving them in a pile on the bedside table. In the morning two little sparrows were asleep in them. They came in at the tiny gap at the top of the window each night while that frozen spell lasted. . . . Our hens were sometimes found dead on their perches, although they were in a good henhouse. Once I climbed on to the roof to look into the tanks to see if they were empty or frozen solid. I fell off, head first, into a deep drift of snow, and was hauled out by a jeering brother. The tanks were frozen.

When your hot water bottle fell out of bed it froze solid on the floor; sometimes the water in it was the only way to have a semi-warm wash in the morning. Like most households in the high country, we kept a fire going in the wash-house under a copper full of water. It was carried in enamel jugs through the house to the bathroom, generally dripping all the way. I remember the tale of a friend calling out to a town guest to take the large container through for his bath; he took a large pot from the range, and tipped out the fowl boiling for dinner!

We had fun with our skates and toboggans, but the short hours of daylight and the long dark nights made the winter seem endless.

We sat around a roaring fire and frizzled in front and froze behind. I liked lamplight, though; it was soft to read and sew by, and gave out warmth.

I remember one wonderful winter sight which we watched for at nights after an electrical storm. Then we wrapped our heads in woolly scarves and the rest of us in heavy rugs, and trudged through the snow to stand and stare at the unearthly beauty of the Aurora Australis, or Southern Lights, those pulsing green, yellow, and pink lights that lifted and shifted over the tops of the white mountains. And, too, the sheer brilliance of the Milky Way. There was the snap and crackle of the frost and the *swoosh* as snow slid off the branches, but our eyes were on the high peaks when the Southern Lights surged up with luminous arcs of light, dimmed, then flashed up again. The colours changed as you watched, sometimes just above the mountains. Sometimes it seemed to us to be miles high, while we children stood awed and silent in the deep quiet of our snow-covered world.

When I grew up I learnt just what housekeeping meant in that awful cold. How would housewives today like to deal with daily problems such as these? We didn't have refrigerators in those days (the 1930s), but during the winter the whole kitchen and pantry were refrigerated, with an oasis of warmth only by the range, a Shacklock coal-burning stove. Butter, to be usable, had to be kept in a covered dish near the range; and a jug of milk, or cups of tea would be stone cold; flour had to be warmed, or it was "dead" and wouldn't rise; chops for breakfast had to be brought in from the safe the night before.

We used a lot of bottled fruit, but sometimes the choice was made for you: you used the bottle that was cracked. Once, I remember, I went off to the dairy for some milk. The jug was in pieces on the shelf, while the milk stood up alone in the shape of the jug. And I remember, too, the extra work caused by having to carry in buckets of hot water for all washing-up, as the pipes were frozen half the time and you didn't dare draw off hot water unless the water was flowing in from the tanks to refill the boiler.

There was a fire in the dining-room which was never out between May and September, and yet, only five feet away on the table, we have lifted the chrysanthemums out of a vase with the water frozen solid round their stems. In the cold spells, as we called them, the periods after a heavy fall of snow when there were hard frosts for weeks on end, everything you did was more of an effort.

You had to remember to take the sheets off the line by 2 pm or they would be like boards and rip into strips if you tried to pull them off. You had to pour a kettle of boiling water gently around carrots in the ground before you could dig them up, or they flew into hundreds of brittle pieces. At that high altitude I sometimes thought, in our dangerous duel between man and his beasts and the conditions that snow and ice could bring about in the Mackenzie Country, that there was only a thin shell between our civilised life and primitive pioneer existence. Our wooden houses were only shells, too; there were few bedrooms with fireplaces, no electric light or heaters so far from towns, and how desperately isolated a homestead was when the telephone lines were broken, until the weary trek over miles of rough country had found and mended the break.

Thinking of all these things, I decided I still like to look at snow: but at fifty miles away, and with the wind behind me.

MISGUIDED

Prudence Gregory

THE TROUBLE WITH being a Girl Guide in my day was that, like almost everything else, the whole thing had been imported from England. It passed completely over my head—and possibly of those in authority—that I was surrounded by the most marvellous scouting country Baden-Powell could have dreamed of.

Smothered in badges and wearing long black stockings, we met in dusty church halls; the Lily of the Valley Patrol with the stage as its own special domain, the Kowhais in the kitchen, and the Rata in the dressing room. Our own gear—Union Jacks, framed copies of the Guide Law and, inevitably, miles of rope—was dragged out each week and draped more or less artistically round the walls, where it competed on pretty unequal terms with the more permanent hall decorations—mostly Sunday School texts and framed copies of the Commandments.

English Guides, on the other hand, had the real thing. Copses, for instance. They were always going on about copses in *The Guide,* a weekly magazine which arrived in bundles every six weeks or so from England. It was our Bible. English Guides also wore "plimsolls" when camping, and usually met in "thickets" on "the common". They instinctively identified the trees and birds around them, and knew exactly how to keep in with the local farmer, who bore no resemblance whatever to his New Zealand counterpart. Also—this caught my imagination greatly—they were forever making plaster casts of animal footprints.

To one bemused New Zealand Guide at any rate, England seemed to be a vast thatched village surrounded by an exciting but somehow accessible countryside, crisscrossed with footprints left by wild creatures of all kinds: squirrels, stoats, weasels, badgers, voles, moles, rabbits and foxes. The Guides, who apparently all lived in the country, mixed plaster of paris with water, stalked silently through the copses till they found foot, pad or claw marks, tipped the plaster into the imprints, let it set, lifted it out again carefully, and at the next meeting on the common everyone said: "Stout work! Caroline's got a spiffing plaster cast of a mole's paw—or pad."

All I could find when we hiked in a dead straight line along the shingle of Napier's Marine Parade Extension, was a cow's foothole, if that is the word, on the riverbank. I got a cast, but it broke on the way out, and anyhow it didn't seem very English. With a vague idea that a pukeko might just be heavy enough to make some sort of dent in the softer mud, I went closer to the water and there saw something much more inviting—Mona's face. She was lying flat out, doing nothing in a most un-Guide-like fashion, while we were fruitlessly trying to read compasses, throw lifelines, or pour plaster of paris into cows' footholes.

Mona was a placid girl, and it wasn't long before most of us were moulding wet plaster of paris on to her face, with the intention of getting a live death-mask to take back to town. She did protest about breathing when she got the hang of what we were up to, but was howled down on the grounds of wrinkling the plaster. I swore we would leave her nostrils unblocked and that if she would only shut her eyes and keep them shut, all would be well. Outnumbered and outranked, she let us pour the plaster all over her eyelids. Perhaps she wasn't usually the centre of so much attention. Once we had her face encased, we left her lying on her back, face up to the sun to dry, while we went off to find due north without a compass.

Mona must have suffered it all as part of being a Guide; she hadn't moved when we returned.

That plaster had certainly set wonderfully. Too late we realised that an unguent of some type would have been wise between face and plaster. I had to knock at it with the handle of my sheath-knife and then prise it up with the blade. All helped—some prising, some reassuring Mona, whose incredible patience had now run out. She was agitated about her eyes and kept picking at them; eventually we had to resign ourselves to a mask with empty eye sockets. Then she got her mouth working, and in the plaster-crumbling recriminations that followed we lost the whole of her jaw.

A good mask was now out of the question; it was just a matter of getting the plaster off Mona's face before she went wild with hysterics. She said she'd felt like the Man in the Iron Mask, all alone and blind; that we were callous; that we hadn't explained; that the hairs on her face made it hurt when the plaster came off.

In the end all I got was a rather jagged piece of nose, and Captain didn't seem to regard that as much of a trophy at our next meeting. Mona's reddened, scratched face attracted far more attention.

Shortly after that, encouraged by *The Guide* once more, I took up woodlore and set about identifying trees—holly, sycamore, and larch.

Suddenly, Napier seemed full of manuka.

W. Jenks.

HOLIDAY ON GRAN'S FARM

David Tolhurst

WHEN I WISH to think of real happiness, I look back on my school holidays spent on my grandmother's farm, high up in the hills.

You came upon the farm suddenly, rounding a bend in the winding, dusty road, and there stood the great white gate, which I never tired of swinging on. It was rather exciting to be left alone, all by yourself, standing before the gate as the service car whined away up the hill. There were things to explore even round the gate—the tall pine trees stretching way up the hill to the right; the large breadbox with its corrugated iron lid; an old, long-abandoned sheep dip; the yards and the concrete water tank alongside, with all sorts of interesting sticks, stones, and pieces of old metal at the bottom, just out of reach, so that instead of retrieving them you had to add a few more oddments—though of course it was strictly forbidden to do so. There, too, you lingered some mornings while waiting for the mail bag and stores to arrive on Newmans' red service car, and siphoned water from the tank, and sailed sticks in its water, and even contemplated jumping in and swimming.

For me the most exciting thing of all was hurrying up the winding drive, stumbling over the rough, white marble chips in a rush to catch a glimpse of the woolshed and the steps leading up to the house itself.

What a romantic house it was, with its long dark hall, off which opened rooms, each with a special fascination of its own and each associated with some part of life on the farm. There was the kitchen, with its large black stove gobbling up wood day after day, and the plate rack above the sink, which eliminated the meticulous drying of dishes, and by the door the telephone, with the stamp book and the toll-call cards which I was sometimes allowed to fill in with a great deal of importance; the boot room, full of old coats and heavy shoes and battered boots for rainy days; the dairy, with its butter churn, milk and cream and eggs—and raspberries, too, if I could find them in the dark.

At the other end of the house stood the austere office, with its marble curios, a model of the sailing ship *Bounty,* shotgun cartridges, and the mailbags. Opening off the office was the store room, its shelves piled high with jars and bottled fruit and boxes—like a shop. In the living room, dark and richly carpeted, the great farm painting on the wall, and the antiquated, crackling and temperamental radio (an Atwater Kent, with three black dials to tune into 2YA Wellington) in a corner beside the fire which sparked so merrily on chilly nights. Close by, all gloom and jumble, lay the sewing room; and further off Gran's bedroom, and the chest of drawers in whose bottom compartment were the treats and toys which appeared on rainy days. And around them all ran the long verandah with its priceless view down to the sea, captured later for posterity by the second mistress of the house, who was a watercolour artist.

Stretching away down the hills was the farm, a pretty thrilling place for a small boy from the city.

A day's mustering over those hills with my uncle was something I always looked forward to in the summer. As I grew older, I believe I really helped him on those excursions, but my dogs certainly never did. Boss, the oldest and craftiest of the dogs, was an unequalled shirker, and after a promising start he would suddenly disappear into the knee-high bracken, to leave me running barefoot in his stead, barking and whistling and shouting at the reluctant sheep. It was hard work battling through the fern and slithering over the outcrops of rock, but there was compensation enough in the wild taste of the fuschia berries; the billy-boiled milkless tea; the mysteries of the old quarry and the overgrown railroad on its way to the sea; the beauty of the tall foxgloves; and the crystal coruscations of the streams among the rich green grass sown by a grandfather whom I never knew. At the end of the day I somehow always had enough energy to climb the last hill leading to home, a warm bath and a late supper under the glare of the gaslight at the big kitchen table.

Then came shearing time, with new faces and the pleasure of sharing my grandmother's hot tea and fresh scones with the men. The woolshed, which before had been so strangely quiet and empty, came to life and was filled with the scramble of sheep on the boards, the click of the press, and the strong smell of greasy wool. Then one day it would suddenly be all over, and after the final baling and loading and the farewell waves and shouts the shed would once again resume its silent shroud.

Most times, however, were times for play; and there, in the loneliest of places, I was never lonely.

Hidden in the trees and the grass was all the world of make-believe and adventure—hide-outs, forts, pirate ships, tiger hunts, wars against the most dangerous of savages and cannibals. When I tired of make-believe there was always the farm itself to explore: hens' nests to discover in long grass and beneath the trees; stoats to trap; dead birds to avoid, for I had an unnatural aversion to them; hares to surprise; and the vegetable garden to raid. I always thought the prickle of the gooseberry and the scratch of the raspberry bush punishment enough for this honest attempt to save the fruit from the birds, but my aunt had other views.

The shed, where lay the derelict winch which once hauled the marble-laden trolleys from the quarry, and the tiny abandoned schoolhouse—they were things of the past, and I hesitated to throw a stone through their cobwebbed windows.

Down below the woolshed grew the gnarled and tangled cherry trees, up which I sometimes crawled at Gran's instruction to gather fruit for jam.

There were the hens to feed, the cows to milk, and the milk itself to separate. On rare occasions we would put cream out on to the woodshed roof in the fond hope that a frosty night might convert it into ice cream—but Mittens the cat would usually have polished it off by the time we were up in the morning.

In those days wealth was of no importance, for we had all we could desire there on the farm. Anyway, there was a scooter at the house, a luxury I never had at my home, so I spent many hours riding it with great fury down the punga-lined path which led to the vegetable garden. I'm not sure that I didn't break that scooter in the end and thereby occasion my mother considerable grief. I certainly remember causing some slight damage to my cousin Frank's toy aeroplane on what turned out to be a rather unpleasant visit to his house.

There was always considerable opposition on the part of the women of the house to my watching the slaughtering of sheep or being allowed out to shoot hares with my uncle, but somehow I won through in the end, like the spoilt child that I was. These exploits took place in the evenings, when nothing we would do could ever spoil the peace of the falling night. There below us grazed the cows, content after the evening's milking, and over the hill shone the red glow of the sun on the dark, tangled native fern.

Home at last, like every good child I would go reluctantly to bed, rather timidly, I expect, down the long dark hall lit by the gas and my flickering candle. And, once tucked up in bed beside the fire that sometimes burned in my room, I would fall asleep with my granny's goodnight kiss lingering on my cheek.

Those were wonderful days, and the beauty of that place has somehow imparted its peace and calm and goodness to the queen of its domain, my dear grandmother.

RIDING TO THE KING COUNTRY

Alex Cassie

THEY WERE HEADING for their new farm in the King Country. It was near Ohura, over a hundred miles away. There was a good load in the gig; my sister Meg, her husband Don, little Jim aged two, and all their luggage.

"You'll let Joe come and visit us in the holidays?" Meg asked. My parents agreed.

The gig wheels spun as they went, and I did a hornpipe while the rest waved goodbye. I had a promise; and my parents wouldn't be allowed to forget it.

Another thing, I owned my own hack now. Spark was his name. He was pretty bright, too. I could borrow a saddle. But even so, I had a struggle: everyone thought I was too young. What rot, there were boys of fourteen smaller than me. I was thirteen years old, and this was 1913.

The first holidays were in mid-winter, only a fortnight. Show time, too. How I got them to agree to my going in winter I don't know, but the time came and Saturday morning found me twenty-five miles on my way, in Egmont Street, riding into Wests' stables, New Plymouth.

Some of the family were in town, having come in by buggy the night before. They shouted me to lunch at the Terminus Hotel and let me go to the show for a short while then shooed me off. I was to make Uruti by night fall.

Going out I met two schoolmates on their way round the sideshows.

"Coming?" they asked.

"No, thanks," I answered. "I'm riding to the King Country."

That made them look!

At the stables I rode out past a long line of stacked gigs; each gig shoved under the one in front. In Devon Street there was a cab-stand, and later a three-horse bus met me by the Red House Hotel.

Behind the bus came a pony-cart with two round faced girls driving it. They wore straw boaters with their school uniforms. At the sight of them Spark arched his neck and snorted. They laughed with bright eyes and lovely teeth.

There were lots of gigs and buggies coming into town for the show, and I saw two motorcars, one of them roaring along, the other on the side of the street with two men's legs underneath. The men always seemed to crawl under when a motor wouldn't go.

At Fitzroy I kept straight on until I found a river in front of me. I asked my way back to the Waiwakaiho bridge. The river looked like the Stony River that I'd crossed in the early morning. I passed a five-horse wagon on the first hill. It nearly caught me up on the flat, and I had to keep trotting to keep ahead.

All this was new country to me. You never knew what was coming round the next bend. Once it was a hay cart. Then a buggy and a pair of spanking blacks, then a three-horse dray. I overtook wagons with manure and slag aboard.

Here I was, riding to the King Country. . . .

At home I went everywhere at a sharp canter. When neighbours heard that I was taking this trip, they each and all advised me to go quietly.

Spark could amble endlessly. He had done nearly a hundred miles a day at times. Now I was letting him take his time while I looked about and enjoyed myself. I was in a new world. I felt like Drake or Columbus. I had no ship but a good horse under me. I had a spur but did not need it often.

At Waitara there was a brand-new bridge . . . and a couple of small steamers in the river.

A cemetery on a corner; and good flat going. I slipped quietly past a mob of sheep. Then I saw a corner with a signpost and just beyond came a four-horse coach trotting along. The people sat up on the box seat by the driver. With the signpost in front it looked just like a Christmas card. No snow, but a wintry look. I rubbed my eyes, but it was real. Next, a mob of shorthorn steers, close packed and well in hand. The drover had businesslike dogs and a lovely hack. Low ridges began showing up ahead. I ambled on.

Soon a rider came cantering up behind me. It was the drover. He asked me how far I was going. "I'm riding to the King Country," I told him.

"You can't make Uruti tonight," said he. "You'll have to move a bit faster to reach Urenui by dark."

We went about a mile at a good canter, then he had to turn off. He was right. It soon began to get dark.

Then I heard a screechy whistle, and a motorcar came roaring past. Its lights showed a mile ahead. After this blinding light the winter darkness seemed all about. It was colder too. I kept a steady trot. It got pretty black but I could see, and trotting kept me warm.

When riders passed they would call, "Goodnight." Gigs had their lights lit.

Then I saw the lights ahead. Yes, Urenui. I trotted along the street, and out of the darkness a heavy drunken sort of voice called:

"Hold on. Eh."

A bit frightened, I spurred on ahead, and soon clattered into the stable yard at the hotel.

The stable-man said: "I'll fix your horse, you're just in time for the dinner gong."

I felt a bit shy asking for things, but I was soon sitting in the warm diningroom enjoying a good meal. People had been at the show. I heard one say: "Scones—goodness, I could make better myself."

Afterwards, I went out to see how Spark was doing. The light over the stable door was shining down. In a stall near the door Spark munched and blew and munched again. The windmill above the stable creaked and spun. I went back into the hotel.

By my bed a notice on the wall said: *Travellers requiring early breakfast should order it the previous evening.* I felt too shy to order anything. I felt tired, and when the hotel lady asked if I'd like to see my room and left me a lighted candle, I thought bed seemed a good idea. She had looked at me kindly and thoughtfully. I wondered afterwards if she thought I was a thirteen-year-old running away from home.

I could hear men talking in the bar, but that seemed to help me drift away into sleep.

I woke to find a lovely morning. I thought I'd better get on the road. I found Spark out with his cover on, waiting with some other horses. I looked in at the feedboxes and the voice of the stable-man called down from above. He told me where the chaff and oats were if I'd like to give Spark a feed.

When I went inside again I found that breakfast wasn't till eight. It wasn't quite seven then. Taking a walk down the street I was hailed by a big Maori man.

"Hi," said he, "want to come fishing?"

Of course I wanted to go fishing, especially when I found that he had a boat.

"Just want two more chap," said he.

I told him that I had to be beyond Kotare by dark.

"You should be away long ago," said he. "Big ride, lots of mud."

Back at the hotel I found the stable-man brushing Spark down. I got breakfast over quickly. I rolled Spark's cover round my pack, strapped it on to the saddle and I was soon away. My meals and bed had cost 3s. 6d.; Spark's items were 3s. 6d. too.

It was a Sunday morning, a bit cold, but lovely passing the trees and gardens.

I kept Spark moving on good going. Later on new metal appeared, smooth, just rolled. Then a roadmen's camp. Further on a chap was trying to catch a draught horse, and I helped him corner it. He put a rope round its neck and jumped on bareback. We yarned as we rode to the camp. He had just come out from Belfast and was working for a Maori contractor. They were rounding up the teams to feed them.

After the camp there was quiet everywhere, nothing moved. Drays and rollers and scoops stood ready for next day's work. Uruti was next and a stretch of new formation. Big signs on the roadside said: *Vote for 6 miles of metal and progress.* I wondered what this meant.

The next sign said: *Mount Messenger.* It wasn't so bad at first, but in the shady bends the mud was belly-deep but with a dry crust on the edge of the road under two feet wide, sort of like a parapet. Spark loved it, ambling along on that parapet above a great bluff with clumps of clay going swishing and diving to the tree-tops below. I wasn't so pleased though. I thought the mud was safer even though it sounded as though Spark's shoes were being sucked off.

There was a strip of burnt papa next; brick-red and hard. That didn't go far, and we were back on mud again. But all around, lovely bush, above and below. Nikau palms and mamakus, in each gully. Yes, lots of birds too—pigeons, kakas, tuis and once a blue wattled crow. I was giving Spark his head. Suddenly he jumped for the parapet and landed with one forefoot waving in mid-air. He shuffled to get his feet planted, but I looked straight below to a lot of tree-tops. He hadn't liked the look below either, but turned to amble along a strip scarcely a foot wide.

I got him down into the mud again. The sticky, clinging clay was making him sweat, but we worked our way slowly—and safely

After a while a rider overtook us. His hack was hard-fed and clipped. He was a stock agent bound for Awakino. He had left Waitara after breakfast—early breakfast. His hard-fed mount was too fast for Spark, who was sweating.

We crossed another stretch of burnt papa, and then could see that Spark would not match the cob, so the man wished me luck and trotted on.

We came soon to the summit and then the tunnel. The going was better on the sunny side. I was enjoying myself now, the bush was beautiful. Suddenly Spark spun around and plunged for home. I pulled

him up and he reared up on his hind legs with his eyes rolling. He swung and snorted as he came down.

I saw then what the trouble was. Round the next bend came round objects with horses' heads in front of each. Rather like big snails. No wonder Spark was scared. Packhorses! There were twenty of them, some had empty saddles. Spark calmed down at the sight of these.

A pleasant voice called: "Got a young one there?"

I laughed. "He was young once," I called back.

The team of packhorses was kept in control by dogs. The packer had them well trained. He sat his mount and chatted to me while his horses picked at the scrub by the roadside. They were in good fettle and very neatly turned out. All but three had saddles up.

The packer himself was neatly dressed. He had a stockwhip with a hunting-crop handle. To me, he seemed like an English gentleman. He said that if I got benighted on the road to pull into any homestead and tell them who I was. And tell them I was riding to the King Country.

"Don't attempt to travel in the dark," warned the packer. "Mount Dampier is much worse than this"—indicating Mount Messenger's bluffs. He had been packing from Tongaporutu. A steamer had called and left several tons of stores. He was delivering the last lot now.

Meeting this pleasant man gave me a happy feeling. I was making better time now. I was soon on the Okau Road. The valley opened ahead. On the sides of the road posts and rails screened sharp gullies. Some of these posts and rails were joined by ropes of vines twisted round them.

I came to Kotare about three o'clock in the afternoon. The little store was open, and there at the hitching post stood Mick, Don's pony. A large collie dog came trotting towards me; it was Lad—dear old Lad, who'd opened the Ohura Trials.

Don was very relieved to see me. We started for home promptly—another thirteen miles.

"When did you eat last? Breakfast, eh?"

Round the next corner he jumped from his horse and got a thermos out of his bag. I'd never seen anything like it. Hot tea and biscuits.

We pressed on till we were hailed while passing a whare where two packhorses grazed outside. The farmer was treating the store-packer boy to a cup of tea—Sunday in the King Country—I was seeing the world.

With this tea, billy tea, we had homemade bread. It had been baked in a camp oven. It was splendid bread, and his raisin "spotty loaf" was scrumptious.

We pushed on.

"Must get over the Grade by dark," said Don.

On one bit where the track was about four feet wide we met some of Don's neighbours: two men, their wives and children. One little chap dived past almost under Spark's legs. His mother grabbed him, then laughed.

"Barney has a horror of the bluffs," she said.

I was glad that we saw Mount Dampier in daylight. We had beautiful glimpses of waterfalls gleaming and sparkling as they leapt, some in shadow and other frothy white in the sunlight.

We were off the Grade by dark. The road was wider now. As we passed homesteads, dogs barked and people came to the lighted doorways to chat a little while.

The frost stung our faces. The mud holes were frozen hard beneath our horses' hooves, sounding like road metal. . . .

We turned in at last where lights gleamed from the windows and we could smell smoke curling upwards from Don's chimneys. Meg called anxiously; I was half a day overdue. Little Jim had stood all afternoon on a stump so that he could watch down the road. He was asleep now.

We entered the pitsawn house and were soon enjoying a piping hot meal and much talk. The packer that I'd met had packed in a piano over the track we'd come over, they told me.

Afterwards, we sat around the widest fireplace I had ever seen. It was grand to be so warm and so comfortable.

I was in the King Country at last.

HOORAY FOR STINKS!

Laurie Swindell

I WENT INTO the shed the other day, and as soon as I opened the door I stopped and sniffed. Apples. Now where were they? I turned my head in one direction and got the heavy smell of paint; another corner, wine. Had one of those corks popped? No, they were all firm. But in a dark corner was the forgotten carton of last season's apples. They were quite sound, but wrinkled and soft and hardly worth peeling and cooking, so I gathered them up for the compost heap and, as I did so, the whole ripe warm smell of them came up in a wave. You know the smell, sweet and winter-appley.

On a wave of memory came the long-forgotten smell of a farm storage loft; the sweet apples, the sneezy chaff, the horsey smell of heavy harness, the sharp polishy tang of the light, elegant gig trappings, the earthy smell of stored potatoes, the unmistakable grain smell, and the pleasant mousey whiff.

I was taken out of today's nondescript-perfumed world to the time when your nose could tell you exactly what was going on.

Walk into my remembered farmhouse, with your eyes shut and your nose open, and there'd be no doubt at all in your mind that this was baking day. Open your eyes, and sure enough the table was laden with biscuits, and wonderful smells were coming from the oven. Another day, and a yeasty smell would tell you it was the day for bread-baking; the acrid smell of fat and caustic coming from the wash-house let you know you'd soon be helping to cut those semi-soft blocks of soap into big bars, then into squares.

And that same smell, honest and soapy, would waft out on clouds of steam, and you'd know it was washing day. Another sniff, and there it was, wet wool, the smell that told you blankets were being soaked in the big tub on the floor, and in a flash you were in, literally, stamping up and down in the warm frothy water, tossing the blankets to clean them, and doing the same job on your feet, because of course you'd ignored your mother's frantic call to be sure to wash your feet *before* you got into the tub.

Wet wool. Another kind too, living; and mixed, strangely enough, with whisky and milk. You didn't need your eyes to tell you there was a cold, half-alive (or half-dead) lamb lying on the warm hearth; warm milk and just a splash of whisky being forced between its silly clenched teeth and running down its woolly chin; a wonderful spring smell that brought you rushing inside to help.

Another smell that made you brake sharply at the door was a sort of breath-catching tang. The house was being polished with mother's own recipe of beeswax and turpentine. You either had to stay outside with your dirty shoes, or, just possibly, be allowed to come in, pull on old bloomers over your dress, and big woollen socks, and run and slide and pull yourself and your sisters around till the long passage was mirror-bright with your work-play efforts.

And there were many more smells; the rare, sweet, powdery perfume that meant mother was getting dressed for a very special occasion; the keroseny smell that meant the lamps had been put out for the night and the whole house would soon be sleeping, everything wonderful and safe.

But what do we smell in a house today? Perfume.

It doesn't matter what's going on, it all smells all the same . . . or almost. The sophisticated nose might be able to distinguish say, spice garden from spring flowers, and lavender from love-mist, but it certainly wouldn't be able to pinpoint the corresponding activity. Now just imagine! You come in the front door of a house and you wonder what's going on. You sniff, and you get perfume. You go to the kitchen, and sure enough the housewife is cooking, but do you get any tantalising odours? Not a bit of it.

The kitchen's been sprayed with air freshener, and any stray savoury steam is being carried quickly outside by a crafty fan that puffs all your wonderful, mouth-watering odours into your neighbour's backyard.

Another day, you stop at the backdoor, and think as you sniff the exotic perfume: "Somebody must be getting ready for a very special appointment." And you come in and find this exotic perfume is wafting from a softly purring washing-machine.

Honest, soapy, washday smell? Never.

Even washing powders now smell like Chanel No. 5. And it's just the same if you're washing floors or polishing furniture: everything simply everything, is perfumed. My nose, at least, is starting to revolt, and longing for something to smell the way it's meant to.

OUR ISLAND OF TUATARAS

Marion Trotter

STEPHENS ISLAND is a bushclad island in Cook Strait. It's many years since I was there, but it seems to me that it was always green, and golden with sunshine, that the birds sang all the time, and happiness was everywhere. When I think about it I can still smell the tang of damp bush and pungent bush flowers.

Sometimes I'd creep out at dawn just to hear the birds. First, just a few soft notes from a solitary bird. It would be answered from across the gully—then another—and another—and another until there was chiming all over the bush. All kinds of birds, hundreds of them in full song. Deep rich notes, soft flutelike notes, some near, some far away, forming waves of song from all corners of the island. Almost as quickly as it had begun the song would die away, and all was silent again, except for the soft rustle of wings in the bush.

Then the sun would rise up out of the sea, pointing long, flowing fingers up the cliff face into the treetops. I would turn and run to meet my father coming from the lighthouse where he'd been tending the light. It was always lit at sunset and put out again at sunrise. The lighthouse is a cylindrical iron tower, painted snowy white and tapering symmetrically forty feet against the sky, grim and lonely like a colossal monument.

All New Zealand lighthouses have different flashing lights, and Stephens Island was two flashes every thirty seconds. It was a revolving light, the most powerful in New Zealand at that time, and on a clear night could be seen for thirty-two miles out to sea. It has been shining there every night for over seventy years.

Stephens Island is high. Its rocky cliffs rise six hundred feet from the sea. There were three keepers who took turns in the lighthouse at night. During the day they were kept busy as all the glass, lenses and brass in the lighthouse were cleaned and polished every day. The big tower, the houses and everything on the island that was paintable had to be painted and kept in order. We had sheep, cows, pigs, and a horse; they always needed attention, and so did the fences and the garden.

Every four months the Government steamer *Hinemoa* called at Stephens Island bringing kerosene for the light, all our groceries, coal, chaff for the animals and everything that was needed. The day the ship came was a holiday. The tiny school closed and the whole population walked two miles to the landing place. Dinghies brought the stores from the ship and they were landed on the rocks with the help of a crane. Then the cases and bags were tied to a trolley and hauled up the steep cliff by a winch drawn by a horse, a mare named Jess, each circle she plodded round drawing the loaded trolley a little nearer.

Sometimes I was allowed to ride on her broad back and this was near to heaven to my eight-year-old mind. Much later, when everything had been carted home, it was so exciting opening up all the boxes. Everything was brought in this way, all our groceries, material to make clothes, Christmas presents, although we needed very few toys as there were so many exciting things to play with on the island. Usually in the middle of the biggest box of groceries, there lay a bag of lollies, for the children.

Captain Bollons always invited us on board the *Hinemoa* for dinner. How we loved it—sitting at the captain's table with so much cutlery we didn't know what to do with it. And the steward waiting on us, handing around dishes of vegetables for us to help ourselves. It was marvellous—like Christmas coming three times a year.

Then there was mailday. A launch would come over and bring our mail from French Pass Post Office every three weeks if it wasn't rough. Great bags of mail and newspapers were landed on the rocks and carried home. Then began the frantic sorting, letters first, parcels, newspapers, and then for the next two days everyone would try to catch up on three-week-old news. No radio in those days.

Sometimes the sea was very rough and we would lie in bed listening to the great waves dashing at the cliffs, going out with a swishing sound only to return a moment later with a tremendous bang. Or the wind howling round the tower or moaning in the bush. We loved it and would tell ghost stories in awed voices, and shiver with the excitement of it all.

Stephens Island is a sanctuary for the tuatara, this strange lizard-like reptile not found in any other country in the world. Tuataras are fearsome-looking, but we children played with them every day and never came to any harm. One lived under our washhouse for years. It would lie for hours sleeping in the sun and looked very old with its grey wrinkled skin, but as tuataras live for more than a hundred years, it is probably still there, under the washhouse.

There is a curious frog, too, which is found only on Stephens Island.

It is the most primitive in the world and an ancestor of all other frog types existing today.

When we were there a man came from the Dominion Museum to study this frog. As I knew every corner of the island, I was allowed to go out with him instead of going to school. What a wonderful week we had. An interesting thing about the Stephens Island frog is that it hasn't got web feet and never sees water. It lives under a pile of stones on the rocky summit of the island. The tadpoles swim about in water inside the egg before they hatch into frogs.

There were many lizards too; brown grass lizards, rock lizards and pretty green ones. All these made wonderful pets, and it was quite the usual thing to keep a lizard up in one's jersey in those days.

The bush is a fascinating place for children to play in. We would start near our back door and see who could go the greatest distance without touching the ground—climbing and swinging from tree to tree like monkeys.

Perhaps the most memorable sight on Stephens Island was the dove petrels. About October they'd arrive in their millions to nest in burrows in the ground. The noise they made at night was deafening; they'd fly about screeching so loudly you'd think their little throats would burst. I believe they come back to the same place to find their own hole, year after year. Only one white egg is laid in each nest and hatches into a grey fluffy chick. The parent birds are away all day searching for fish and return at night to regurgitate it for the young. After three months, when the chick is larger than the parent, it is left to feather and fend for itself.

The brilliant beam of the lighthouse attracts the birds and many die through striking the panes in flight. It was quite a common sight to see a dozen or so dead birds at the foot of the tower—to say nothing of the mess they made of the glass.

When we left after more than five years on Stephens Island and returned to so-called "civilisation", I hated the noise, the hustle and bustle.

I hated the chattering of people, the sound of hurrying feet, the noise of cars and trains, and longed for the solitude of our island again. I hated the hard city streets, too, and the new shoes and stockings that were bought for me. I longed to kick off the shiny patent leather and run barefoot on the thick carpet of brown leaves in the bush; or to feel the dew between my toes as I ran to meet my father in the early morning as he came from the lighthouse.

THE MAILCART GOES THROUGH

G. J. Nutsford

THIS IS SATURDAY, the day my dreams and fantasies have full sway; no school, no sudden swish of the dominie's cane and his often repeated saying, "Dreaming again, boy? Get on with it."

Lying in the cold dark waiting for the ringing of the alarm clock, my excitement rising to choking proportions, I picture the day ahead, the people we will meet, weaving romantic stories round each. When at last the strident ringing of the alarm clock echoes through the quiet house, I get up, fumbling in the dark for shirt and trousers, and carrying my boots down the still dark hall, brave but secretly afraid of the dark.

My father follows the trade of a wheelwright, a trade which will die with the passing of the wooden-wheeled horsedrawn vehicles. We live in the eastern district of Southland, in the little settlement of Wyndham. And every Saturday, perhaps to earn some extra money to feed so many mouths (we are a family of nine), my father drives a mailcart for the Post Office, and I have been riding with him on most of these trips for the past few months.

As my father and I leave home and make our way to the Post Office Depot at six-thirty, the frost is settling down through the darkness and the grass already has a crispness and crackle under foot.

I shiver, huddling into my overcoat and stamping my feet on the gravel road. As we cross the rail tracks their dull gleam in the faint starlight points in parallel lines towards Edendale and, to me, that unknown world beyond.

At the depot we pass through a wide opening between two long, low sheds, open in front to house the four-wheeled mailcarts. The gravelled yard has stables on two other sides.

Here indeed is a scene of exciting but orderly chaos: the pungent smell of dung and sweat and dust; the pale light struggling through the smoked glass of the two lanterns hanging on the stable wall doing little to dispel the darkness; the large open door showing the yellow light of several lanterns inside; the movement of horses, the occasional whinny or the stamping of a shod hoof on the hard ground of the yard;

the quiet coaxing voices of the wranglers or stablehands, with now and then an aggrieved curse; the sound of wheels moving through gravel, and the harsh squeal of a wooden brake-shoe hard on a wheel.

Some drivers have already left on their runs, others stand talking or stowing their mail and papers in their carts. As the horses of each are harnessed, away they go, leaving the yard at a gallop.

The horses are almost wild, half-broken, and my father is convinced that some of them have never before seen a cart. They can't be trusted to stand unless the brake is hard on and the reins tied to a wheel; they would leave you deserted and stranded miles from town.

The stars are already paling, and the lanterns, having as it were given up a losing battle, are now surrounded by a faint halo.

"Is that you, Bill?" I recognise Bob's voice. "Are you ready?"

"Good morning, Bob," says my father. "Bring them out."

I have always admired these horse wranglers—they seem fearless of the most vicious brutes; although their language is usually lurid, as a rule they speak quietly and patiently to their charges, even though they have been feeding and grooming them from the very early hours.

This morning we are being attended by the wranglers called Bob and Charlie.

Watching the stable door, I see Charlie come out first leading Lucy, a small black mare. Lucy is perhaps the one horse with no vice. Liked by drivers and handlers alike, she has a habit when you stroke her of pressing against you and rubbing her head gently up and down your coat.

As she passes the lantern just inside the stable door, its feeble light shows up her well-groomed coat like black diamonds, and she greets the fresh morning air with a snort and a toss of her head in approval.

Bob comes out leading a big roan on a very short rein. As he nears the door the horse lays back his ears and charges through. Bob, evidently expecting this, is quickly to one side and with a jerk brings him round in a tight circle.

Charlie holds both horses while Bob fastens the traces.

"Up you get, boy," says my father, and he follows me over the wheel.

Sitting on the high seat, I grasp the iron bracket on the side.

My father takes the reins from Bob, looks down at him and says: "Yes, all right Charlie—let 'em go."

At the movement of Charlie's hand the big roan gives a tremendous snort, springs forward at a full gallop, and is pulled up so suddenly by the traces and the hard-gripping brakes that he staggers and almost falls.

Bob's shout of laughter is drowned in the pounding of hooves and the clashing of iron-tyred wheels on the gravel as the brakes are released and both horses take off again at breakneck speed.

As we turn right, through the two sheds, the wooden shoes of the brakes grip harshly. The wheels momentarily stop turning and go into a juddering slide, the wheel on my side coming so close to the wall that I call out in fright, but the brake is thrown off at exactly the right moment and we swing out into the street.

Lucy, with her shoulder boring into the guide pole between the two horses, is using her experience to force the roan into the turn.

The roan, already wet round the flanks, great gobs of saliva flying from his open mouth, pounds down the road, taking the whole weight of the mailcart, which is not much in any case. Lucy, with traces hanging loose, keeps pace alongside him, as usual enjoying the run.

Gorse hedges, an odd cow or two, take shape in the lightening dawn as we begin climbing the lower slope of Young's Hill. The roan starts to flag a little and my father, usually quiet and patient with the horses, takes the whip and cuts him smartly on the flanks.

Startled, the roan's flying hooves strike fire from some hidden flint rock in the road; my look of astonishment at my father is met with "Just to keep him going for a bit. It'll take some of the dirt out of him."

Our pace as we pass Young's gate at the first bend has slowed to a trot, and I throw out my first bundle of newspapers. My hand holding the iron stay of the seat is aching now with cold. My feet are like blocks of ice.

The slope is steeper and the horses slow to a walk. Climbing over the back of the seat I drop to the road and walk too the rest of the way up the hill. My father has already sprung over the wheel on his side, controlling the horses as he goes.

It's now daylight, and as I plod along I'm in a small cloud of steam from the horses' breath, smelling faintly of oats and the sweetness of molasses.

Near the top of the hill, where more papers and mail have to be put in a box, my father climbs aboard and hands the parcels to me. As I clamber back on to the seat, I look back over the valley.

Laid out in seemingly neat squares, with its many types of hedges and fences, I see magnified, the old quilt my mother had made out of patches for my bed. Almost directly below is a sweeping bend of the Mimahau River, its waters a black line as though drawn with black paint, while far to the west, where the sunshine is creeping out on to the plain, a line of green marks the willow-covered banks of the Mataura.

The brittle-brightness of the sun is tossed back and forth from myriads of inch-long icicles hanging from the gorse hedges. As I climb down to put more papers and letters into a box, I wonder who has spun this silvery frosted web across the opening of the box, this web I am so loth to break; surely not a common spider. And look at that delicate pattern imprinted on the ice-covered puddle at the foot of the gatepost. . . .

Rabbiters with their packs of dogs meet us at odd places; some with beards, dirty, torn clothes and obviously unwashed, their language so coarse and obscene that I can't look at them for shame and my father, who can swear with the best of them, remonstrates, "Put a sock in it, Curly!"

One big bearded man with a mop of untidy grey hair chews tobacco, and the overflow has stained a neat track down his greying beard. His conversation, mainly about prices for rabbit skins, is punctuated by

squirting tobacco juice at any dog coming within range. I remark to my father as we drive on that this old rabbiter is a pretty good shot with his tobacco juice.

Some of these men, although alone for months on end, keep themselves respectable, and usually have a parcel of books from the librarian at the Athenaeum. We deliver these, and pick up parcels of books to be returned.

Approaching Waimahaka, I clamber into the back and come up with something special: a large, precious box for Mrs Byers.

"What do you think it is, Pop?"

"Only one thing it could be, boy. A new hat."

The house is about half a mile from the road over a very muddy track. But Mrs Byers is out and waiting for us at the road gate. She's dressed in a pair of man's trousers, with big boots and leather leggings, a tartan shirt with sleeves rolled to the elbows, showing very heavy forearms with reddened skin.

"How are you, Bill?" she bellows. "Have you got my mail order?"

As the horses stop, I hand the box down to her.

"Let's have a look at it." Her voice has lost none of its power, and Lucy the horse looks round in curiosity.

Putting the box down on some clean grass, Mrs Byers tugs the string off impatiently, and drags out—a hat!

Black. Wide-brimmed, decorated with an enormous bunch of artificial flowers. A white ribbon dangling in two long streamers down the back.

"What do you think of it, Bill?" Not waiting for a reply, she pulls the new hat on to her head, her round face all smiles.

"Lovely. Lovely, Mrs Byers; but we must be off—we're running late."

My father's voice sounds as though he has caught a cold, and he whips up the horses so suddenly I'm nearly tipped out over the tailboard. Some distance up the road my father explodes into a roar of laughter.

The next stop is Warakiki, a changing station, where we enjoy a hot meal and are given a change of horses. Other mailcarts and some rabbit carts stand around the station. My father knows some of the men, and over the hot stew and potatoes they exchange news.

Two drivers get into an argument about riding horses, and one challenges the other to a match. We'll see a bit of fun, all right!

After the meal the wranglers bring out two horses into a small holding paddock. The men draw straws for first ride. The wranglers hold the horses, which have open bridles and no saddles.

The first man is helped on to his mount and I think nothing is going to happen. The horse stands perfectly still, then with an angry squeal rears and swings round as if dancing. With a movement almost too quick to follow, its tail is up in the air where a moment ago its head was. Its four feet on the ground again, it spins round and round, jumping high and landing stiff-legged, with head between front legs. The rider's nose begins to bleed, then suddenly the horse swings sharply round as he jumps. Two turns like this, and the rider is unseated and falls with a thump on his back.

He doesn't seem hurt, though, and is on his feet, wiping his nose on his shirtsleeve.

The second man, a bit white in the face, goes out bravely to his horse.

His ride is soon over. The horse, probaby more scared than his rider, bolts straight at the gate. A few feet from it he stiffens all four legs and comes to a jolting stop. The men roar with glee, and stamp and wave and cheer as the rider clears the gate and sprawls in the dust.

But it's high time we were on the move again with the mailcart. Outward mail which had been collected at the station is stowed in the cart, along with grocery orders, books to be exchanged, and messages for friends or relations in the town.

The carts pull out of the yard to a chorus of shouted farewells, the cracking of whips, barking of dogs, and wild whoops of fun as each driver careers through the gate to the road, showing his horsemanship and the fleetness of his charges. We are now driving two bays.

The most exciting part of the run home is the open creeks we have to ford, but it's frightening when we ford a part of the Wyndham River and the water comes swirling up to the floorboards. The horses snort and seem to be swimming at one stage. Although they are tired when we reach the outskirts of Wyndham, they prick their ears and take Main Street at a pretty fast clip.

Night comes early to Southland in winter. We hand over the two tired horses to the wranglers. We deliver our mail and orders to the office. And, side by side, my father and I make our way home through the gathering dusk.

ESSAY ON A RAT

Margaret Gibson

IT WAS A DAY for being outside, anyway, and there we were, shut up in our one-room school till three o'clock. We were all lazy—the teacher too, I'd imagine—with the windows open and letting in the hot sunshine and the heat that beat up from the playground.

We'd been given essays to write and had got to the stage of reading our efforts to the rest of the school; whether they listened was another matter.

I remember only one essay. Although I don't recall the exact words, I'm sure the first four words were "I am a rat". When we wrote essays, whatever the subject, we never missed the chance of making ourselves the heroes or heroines.

The rat was an outdoor type in this essay, just as I was myself in my schooldays. I sat at my desk like the others, but no longer drowsing in the schoolroom; I was away with the rat.

The rat wandered about the places we explored so thoroughly after school and on Saturdays. I saw him go along the creek where the water slipped up into the long soft grasses, where the edges were so shallow it was hard to know where dry land stopped and water started if you wanted to keep your feet dry.

I was with him as he stared into the deep dark-brown places under the roots of flax bushes, where we believed eels lived. Together we warmed ourselves on the baked ground at the bottom of a honey-scented gorse bush.

Then the rat shot up the trunk of a willow tree leaning over the creek, and I went after him, light as a leaf in the breeze, up among the twigs where I'd never been able to go before. He ate a couple of thrush's eggs from a nest, but I did not share an egg with him.

Down we scampered again and headed away across the paddocks, sometimes following the roughness of a sheep track and sometimes wandering along the rat's private ways, back to where he lived in the stable belonging to the father of the essay-writer.

There were stalls along each side of the stable, doorways at each end, and down the full length a wide stone-cobbled passageway.

It was the most daring thing in the world to walk down the centre of this passage when the horses were in the stable, between the double row of rear ends of large farm horses snuffing at the chaff in their feeding boxes and jinking their harness. There was always the possibility that a horse on each side would at the same moment stretch a hind leg out, and you'd die, clamped between the two crushing hooves.

The rat and I went into the stable. The rat ran along under the feed boxes, up the wall, along a rafter, and into the loft which was home to him.

So the essay ended, and I found myself back in school; and to this day I remember the shock at finding myself there. Words had never done such a thing to me before. Words were reading and spelling and, less noticeably, talking; and from the day when the words, "The cat sat on the mat"—helpfully illustrated—made sense to me, I'd liked them.

But to find words could pick me up and carry me off like that was something new, and a great shock to be absorbed a little at a time. Other people's words could waft me off from reality . . . so . . . perhaps I could write words that would do that to other people.

I still meet today the girl who wrote that essay. I've never tried to discuss it with her because I'm sure she's forgotten about it long ago.

But of all the essays I heard or wrote in my schooldays, that's the only one I remember. Even writing about it now has recalled the atmosphere of that hot little schoolroom, the children lackadaisical with the warmth, me more so than most, probably, until the story began and I was swept away—carried completely away, until I heard the words which almost certainly ended the essay: "He arrived home, tired but happy."

On The Job

WHEN THE SCHOOL COMMITTEE MEETS

Patricia Russell

"THE TIME has come," the chairman said, "to talk of many things,
Of Catchment Boards, dog dosing, and profits that have wings,
Of all the many problems that distress the farmer so,
And the school committee business we shall fit in as we go."

These words had scarce been uttered, when a voice in anger cried,
"That nincompoop who doses dogs, by God he's got a hide,
Here's lambing time upon me and he wants to shoot my bitch!
I'd like to take a shot at him, and chuck him in a ditch."

"Don't talk to me of ditches," another fellow roared,
"They're a lot of cranky halfwits on that flaming Catchment Board.
They don't know B from blazes, as to how a ditch should run,
And they'll have us cockies bankrupt soon, if something isn't done."

Then up spoke one brave fellow, and in accents loud and clear
Said "I'd like to see the trees around the playground trimmed this year,
Suppose we scout around a bit, and gather up a mob,
It wouldn't take us long if we got stuck into the job."

You could have heard a pin drop. Not a member made a sound,
But everyone who sat there glanced uneasily around,
And then they spoke up with one voice—they'd willingly be there,
Except that not a single one had any time to spare.

To save the situation and put everyone at ease,
One chap commenced to tell a yarn he knew was sure to please.
Then all the rest guffawed, and at the height of all the fun
A voice said: "Yeah, well that's not bad, but have you heard this one?"

Then the chairman called for order, and when all was quiet again
He volunteered the latest information on his "pain";
He'd been to see the doctor, who had prodded him about,
It'd turned out to be gallstones——he'd have to get 'em out.

"Me brother had his out last year," the secretary said,
"He'd been crook as hell for months, spent half his time in bed,
But since his operation he's a fitter man, by far,
And he's got the gallstones on the mantel, pickled, in a jar."

The chairman muttered coldly: "Well I'm glad your brother's fine,
But the doctor said he'd never seen a case as bad as mine.
Now suppose you read the minutes, and please try to get them right,
I don't feel up to sitting in this bony chair all night."

They were halfway through the minutes when old Bert said "Dear oh
  dear,
How could I second all these things? I wasn't even here."
The secretary grinned and murmured, "Now, don't make a fuss,
If you weren't here last meeting Bert, then that makes two of us!

"I say that what is written here is absolutely right.
If anybody disagrees, then I'll resign, tonight.
I didn't want this lousy job, but no one else would take it,
So if you've some wisecrack to make—just go ahead, and make it."

"Whose turn to make the supper?" asked a quiet voice from the rear.
"I'll make it," said the teacher, "carry on as though I'm here.
When you've got the minutes over there are things that I must mention,
Important school affairs I'd like to bring to your attention."

Then Bert said, "This is thirsty work, my throat's as dry as sand,"
He ducked outside, and soon was back, a peter in his hand.
The secretary took a swig, and gurgled, "That's the stuff,
And as for these damned minutes, I think I've read enough."

They told a few more stories while the teacher made the supper,
And with the peter empty, settled in to have a "cuppa".
Then the chairman thumped the table, sloshed the tea and broke a cup,
And to the teacher said, "Now Jack, what's this you want brought up?"

Jack cleared his throat: "There's lots of new equipment that we need,
Some sporting gear, some books, new ones, called *Teach your Child to
  Read,*
A year ago you promised me new lino for the floor,
And cupboards for the footy boots, built in the corridor."

The chairman glanced up at the clock. "By gosh it's getting late,
I'm sorry Jack, it looks as though your needs will have to wait.
I tell you what, I've got a brainwave that should save the day,
We'll bring the matter up next Thursday, at the PTA.

"We'll kid to them, and maybe they would run a Bring and Buy,
It won't hurt them to do some work. Now fellows, I must fly."
"Oh well, that's that," the others said. "We may as well go too,
We've dealt with the important things, there's nothing else to do."

And so they jumped into their cars, and took off with a roar.
The teacher emptied ashtrays, washed the cups and mopped the floor.
And as he worked he muttered, "By the Powers, there's nothing beats,
A night in Kikamukau, when the School Committee meets!"

WHEN THE CAGE FELL DOWN THE MINE

Dick Carson

IT'S ALL completely overgrown and forgotten today, down the West Coast, five miles in from Reefton. . . .

The winter of 1916 found me working as braceman for this large goldmining concern. The braceman is in charge of everything that goes down a shaft: men, tools, timber, explosives. He wears no fancy uniform, no brass buttons, but nevertheless his rule is supreme in his own domain.

This became clear enough one day when the chief superintendent, his secretary, the management, and other officials came on an inspection tour. They packed themselves into the cage along with their gear, and were apparently amazed when I refused to lower them. I insisted that their gear was removed from the cage, to be sent down at my discretion later on in an empty cage. I never did learn if the manager was as abashed as he appeared, or if actually he was laughing up his sleeve. Anyhow, this was my job, the braceman, at the top of the mine. My opposite number working the chambers far below was called the chamberman, and was in charge of everything sent up the shaft.

Unfortunately, in this shaft which was a little less than 2,500 feet deep, there was no telephone or other means of communication with those people working down below. The blacksmith's fire was about twenty-five feet away from the top of the shaft, and when everything was covered in snow, we were glad to get around it whenever the chance offered.

On this particular day we had been sending down timber, and when it was all disposed of the chamberman came up to the surface to sit around the blacksmith's fire to warm his hands and enjoy a snack and a good hot cup of tea. When he went below again, down to his dark domain at the bottom of the mine, he told me that he was sending up a full load of old waterpipes. So far, so good. I settled comfortably round the fire while the cage was being loaded with these old pipes 2,000 feet below. When I heard the pull up signal, I casually strolled to the mouth of the shaft. I heard a frightful din and instinctively grabbed the knocker line, intending to signal the winder to stop pulling. Just as quickly, I realised that there were no men in the cage and decided to let the winder pull right through.

What a mess came up to my view: the heavy steel gates were torn off, the springs of the patent safety-grippers were ruined, the cage was bent, and every single waterpipe had disappeared. This cage was definitely out of commission. What a mess the shaft must be in, tangled through and through with bent and distorted waterpipes. There was no telephone, remember, and no other method of communication with those below.

I conferred with the winder, and decided we'd try to vibrate the cage without lifting it more than a couple of inches. No response came from below, so we reckoned the cage was not being worked at that moment. Next, the cage was raised about eight inches and dropped, then about eighteen inches and dropped. We had to have that cage, and no man could be allowed to come up that shaft in its dangerous condition. At last the chamberman got the message and rang the cage to the surface.

When the cage arrived I examined it carefully, particularly the open jaws of the safety-grippers, with their four rows of formidable teeth. Then I had to go below for the chamberman. Together we would have to check over 4,000 feet of shaft (two compartments) before any miner could get to his home that night. So down below I had to go. I stepped on the cage and rang away—a ring which meant "Danger, lower very slowly". I was simply aghast when the cage immediately began to plunge at speed.

I thought: "He's missed his clutch," but a moment later I knew: *the winder had lost control.*

I believe most adults have at times wondered how they would die. I knew how it was to be with me. I did not fear death, but I have always hated pain.

"What a lonely death," I thought. "All alone in the dark: four stone walls around me, steel above and steel below."

It wasn't the thought of death that troubled me, but the idea of my body being pierced by the protruding ends of those waterpipes. And all the time the cage was plunging down in the darkness.

I grabbed the centre-bar and hung by my hands, my knees almost touching the floor of the cage. If I hung low I felt there would be less chance of a pipe piercing my face, my throat or my chest.

Then *crash,* as steel crashed through steel—the noise was terrific—sparks lit up the cage momentarily, then *boom, boom, boom,* as broken pipes sped on their way to the sump at the bottom of the shaft, so far

below. Another *crash*—this time the safety grippers began their rocking, with their dull *zig a zig a zig—zig a zig a zig a zig—zig a zig a zig.* But alas, those life-saving jaws could not close. Then the very danger I was out to defeat overplayed its hand, and jammed the cage sufficiently to slow it up a little. Immediately the safety grippers dug in with a vice-like grip. The long *iss iss iss,* almost like escaping steam, told of how those teeth were clinging to the wooden skids. This easing up of the speed of the cage gave the winder his opportunity: he managed to slip in his clutch.

There was a terrific echoing shattering noise as if the whole cage had been shattered. But I knew it was the sudden forced release of those terrific clinging jaws. The weight of the cage had once again been taken up by the haulage rope: everything was back under control.

The cage had now slowed up considerably as it passed No. 9 level, gradually slowing to come to rest at No. 10.

Stepping off the cage, I said to the chamberman: "Did you hear anything as I came down?"

"Hell," he exclaimed, "it's a wonder you're alive."

We both knew the job before us now. We boarded the cage and gave the signal, "Men on board—pull up." Now you understand, when a cage is being pulled up and the other cage is acting as a back balance going down, it is a steady even pull. Pulling singly, however as we were now, with a huge drum measuring 14½ feet in diameter, at the end of every piston stroke there was a jerk. We were in for a rough trip.

All of a sudden the jerking stopped. We could feel an unusual strain on the rope. A continual banging began on top of the roof of the cage. There was silence between us, but we were both thinking the same.

Sure enough, my mate echoed my own thoughts when he said, "If anything snaps, I hope the grippers hold".

When we arrived on the surface, four pale faces were peering down the shaft. They had heard the banging echoing in the shaft, and knew men were in the cage, but they could not do a thing about it.

Upon examination, we discovered that a long waterpipe had wedged itself down between the skids and the guides and was bent over the top of the cage. So every time a wall plate was passed, the bent pipe was jammed against the top of the cage. Here was a job indeed.

When we righted it, we stepped on the cage, gave a "4—4—2" ring, and began to descend slowly to make our examination.

I knelt on the floor, shading my candle in an attempt to see any loose material which might be dangerous. The chamberman stood showing his candle into the empty chamber of the shaft, his fingers sliding loosely down the signal line, ready to signal "Stop" at the least sign of danger.

Time and time again we had to stop the cage, crawl into the vacant chamber and, standing on the slippery greasy wall plates, wriggle bent and broken waterpipes so as to load them in the cage and tie them to make them secure.

One slip meant goodbye, but it had to be done before the men below could get out and home. We might have had to work until eight o'clock next morning, but luckily we got the men out by midnight.

To make doubly certain that the shaft was safe, we sent up a number of skips of quartz.

The chamberman came up with the last load of men. As he stepped off the cage I pulled down the safety gate.

Another shift was over.

IN THE HOPS

Doris Parker

A HOP GARDEN ready to be harvested is a lovely sight, with its even rows of tall green vines. From the top almost to the bottom hang graceful branches of hops, rather like little pinecones, a lighter shade of green than the leaves under which they hang. It seems a great pity to destroy such beauty, but time and weather would bring destruction if we didn't, and anyhow the grower wants his crop harvested, and the pickers wish to earn money. And so, worked out in systematic patches, down come the vines, cut by the ruthless cat—a reaphook on a long handle—of the stringer.

To be a successful stringer you've absolutely got to have good muscles, speed, tact, good temper, unfailing humour, and the ability to handle people in more ways than one. Our youthful but experienced stringer filled the bill very well, passing up and down the line of bins, answering the call of "String", placing the vines high on the end of the bin, or down at the foot as wanted. He's quite popular when the hops are good, and from round about you hear pleased remarks like: "What a beautiful string!" "Aren't these hops lovely—look at the length of them." "We've got a good crop this year."

But even the best gardens have less good patches, and then the stringer isn't nearly the good fellow he was, and the tune changes to: "What do you call that thing?" "Look at this bushy brute—I've a good mind to bury it." "I'll never come to the end of this beastly thing." "Hey, look what you've given me. Don't put that rubbish in my bin!"

The hops are picked into a large bin, of canvas or scrim, on a wooden frame. Most of the time the pickers are shaded from the heat of the sun in February or March by the tall vines, and it's possible to work in comfort. The wise and experienced bring boxes to sit on when standing becomes too tiring. Failing a box, try the edge of the bin, provided it's not too high, well padded with a folded coat or hop sack; that makes a comfy seat for a time. Fingers work busily and tongues wag merrily to the accompaniment of falling hops.

The cry of "The billy's boiling" halfway through the morning and afternoon brings great joy to all hearts, I can tell you.

Towards the end of the morning and also as the afternoon wears on, the hops are measured from the bins into large bags—hop-pickers are paid by the bushel, and over the years the price paid to the pickers has increased from 2d to today's 12 cents a bushel. The bags full of freshly-picked hops are heaved on to the truck, and away they go to the old kiln, which has dried the hops since the early days. The aroma of those drying hops is a memory which will remain with us throughout our lives.

A hop garden would seem a strange place without children. For many years the schools round Motueka and Riwaka districts had a month's holiday so that the children with their mothers could harvest the hop crop. Many families earned their winter clothes in this way. Some years ago teachers and the Education Board decided that the month's holiday so soon after the Christmas ones was not in the best interests of the children's education; and some objected to what they called "child labour" being used to harvest the hop crop. So the holiday was abolished, to the regret of many children and growers.

The time-honoured custom of "putting in the bin" goes on a bit during the season, and is really in its stride on the last day—into the bin you go, regardless of whether you're a child or a grandma. One little lady advised her niece, new to hop-garden ways, not to struggle or protest, but to yield quietly and gracefully. She herself gave a practical demonstration a moment later uttering squawks and protests when suddenly seized. The niece commented several times on people who gave advice, during the rest of a merry and vigorous day.

Well, eventually all strings were picked, bins cleaned and measured, the custom of "putting in the bin" observed, and goodbyes were said. Hop-picking was over once more, and what a desolation in place of that once-beautiful garden, with dead and dying vines spread and stripped and trampled all over the ground. But soon the rubbish would be gathered and burnt, and later the ground ploughed and left to fallow for the winter.

Our Nelson Province hops are used chiefly in brewing beer to give it a flavour. Hops have no alcoholic qualities; just flavour. They are used in medicine too, and some people have found that a hop pillow is good for getting you off to sleep. And, by the way, apart from satisfying our New Zealand brewers, we export hops to Ireland and to Australia, if we have a surplus.

In the early days bakers, and housewives too who baked their own bread, made the necessary yeast from hops and potatoes, and bottled the stuff down. Many a cork insecurely tied down shot up to the ceiling followed by the yeast when it began to work.

The early settlers in the old sailing ships brought hop sets, as the plants are called, with them, chiefly from Kent, but also from Worcestershire and Sussex. The plants were wrapped in wet straw to keep them alive. Gardens were established in Nelson as soon as ground could be cleared and planted.

Hop growing spread to the Waimeas, the Moutere, Motueka, Riwaka and, for a time, Takaka had one or two hop gardens. The oldest garden in the district belongs to the Holland family of Waimea West and has been going strong since the beginning of the century. Mac Inglis of Motueka has the largest garden in the whole hop-growing district, containing thirty-six acres.

Several varieties of hops were grown in the early days. One variety called Bumford had a reddish vine and grew small hard hops which rattled into the bushel measure and proved most unpopular with the pickers, who like lightweight hops to make good scores. The growers didn't like them much, either. A greener hop called "The Grape" was grown for years, and then almost entirely gave way to the Californian variety, a much heavier cropper. So from the English ones we switched to American hops. Unfortunately some years ago the Californians became badly affected by black root-rot, and are today replaced by three other kinds called "First Choice", "Smooth Cone" and "Calicross" which are tolerant to the disease.

There's another troublesome disease which attacks the hop plant, especially in a dry season, and that's Red Spider. Dr Rudolf Roborgh, of the Hop Research Station in Riwaka, who is responsible for the new varieties of hop, hopes to breed new strains soon which will resist both black root-rot and red spider. Things like that mean a lot to the grower, and to the industry.

In early days and until about sixty years ago, the hop crop was grown on tall manuka or willow poles planted in even rows. Hops growing up poles meant a lot of hard work. At picking time the pole was loosened by a gadget called a dog—so there was a dog for loosening poles and a cat for cutting down the vines—and the pole was pulled out of the ground and placed on the cross pieces of the bin. When finished, the pole was turned off the bin by the picker, and by the end of the day quite a pile of poles had accumulated by each bin. The vines were trained around the poles and tied with strips of bulrush or soft string. All pretty laborious it seems now.

All this changed with a vengeance when today's post-and-wire gardens came in to make the labour of hop growing and handling much easier. Mind you, the cost in the beginning was heavy, but after that, hop twine or string was the only cost each year. To hold the weight of the crop the garden needs to be very strongly constructed, with plenty of great sturdy posts, often made of black birch, spaced regularly thoughout the garden. The height of the wires carrying the crop is from thirteen to sixteen feet.

The rows of hop plants, called hills, are six feet six inches apart each way, 1,000 hills to the acre. If the crop is good, there should be three bushels to the hill.

A good picker, generally a woman, although some men do well too, makes a good score during the day of forty, fifty or sixty bushels or even more. Some people by great effort and working long hours have picked from eighty to a hundred bushels in a day, but such scores aren't kept up, and they're pretty rare. Most of us have to be content with from twenty to thirty bushels daily—that earns you about $2.40 to $3.60 a day.

A really good picker is not only a good scorer and good tempered when things go well, but goodnaturedly tolerant when things get a bit scratchy. It's often said that your true character shows up in a hop garden, and some people certainly are growlers, while many never complain except in a goodnatured way occasionally. The usual picking time is from 8 am to 5 pm with breaks for morning and afternoon tea and lunch. Some people go early and pick late to increase their scores.

The first hop-picking machine was imported in 1951 and installed at the Research Station in Riwaka. The idea caught on. A few growers imported machines, and each year more did so, until by 1969 only one or two gardens were hand-picked. In another year, it may all be done by machine, and another colourful New Zealand countryside scene will have come to an end.

MUTTONBIRDER

Tom West

On a wet winter weekend, what better than a roaring fire and a kit of delicious Stewart Island muttonbirds! And here's the veteran Tom West returned from a successful expedition to the Muttonbird Islands, in the Far South:

I'M A FISHERMAN, and when the weather suits I fish. When it doesn't, if it's the season—that is, April and May—I try to collect muttonbirds. And if it doesn't suit that either, I hunt for a nice quiet anchorage and wait there till the weather improves.

In the first part of the season you take the young birds from the burrows: later, from outside the burrows. You must have suitable clothing, gloves attached to your jacket, and in our case—for not all muttonbirders use the same method—a cane about two and a half feet long with a hook on the end. The hook has no barb. You feel in this burrow where the muttonbird is: you can see by the look of the burrow if it's occupied or not. In any case, you are supposed by regulations to try every hole and clean it out.

After you've tried a hole and found a muttonbird in it, if it's not within an arm's length you use your hook. If you can't hook it out, you note the direction the hole is leading and you cut downwards into the burrow, say a foot deep, perhaps more, but sometimes quite shallow, only about three inches.

You then usually can reach the young bird, take it out, and carefully re-plug the hole you made down into the muttonbird burrow. It has to be watertight. There's a fern growing on most muttonbird islands used for this purpose, and the plug is called "pru". It is quite easily removed by hand, but if you miss a year the roots grow though the hole and become part of the ground; then you have to dig it out. Thanks to these prus, when the ground has been used by a good muttonbirder for some time it's very easy work, very quick. You don't have to do any—well—*research* looking for the nest.

It seems to me the mother bird must have wonderful instincts when she selects the place for her nest, for it's always where the young ones will have access to the sea without any difficulty: nearly always where there's a downhill run to the sea or the cliff, but sometimes where they can take off from a high part of the muttonbird island. There are several take-off places on high islands like South Cape.

The mother birds go away in the morning, and come ashore in the evening, just about dusk as a rule, when they're feeding the young ones. They stay there all night, and just before dawn they come out and give their call—more and more and more—and each attracts another one, and they all congregate before they take off. They don't come straight out of the hole and go over the side and into the water. For some reason or other they gather in big numbers, then all of a sudden, almost as at a signal, they take off in thousands. I've been out there; it's a sight worth seeing. Thousands and thousands of them jostling one another down this runway, which is polished with their feet.

The second half of the season starts about mid-April when the young birds are being fed less by their mother. The mother reduces the feeding, the young ones get restless and move about, and come out of their holes. The mother, as I understand it, encourages the young one out. As she leaves in the morning, she'll often be seen sitting at the mouth of the hole calling out to the young bird, with its head sticking out. Later, they become quite keen to quit the hole at dark.

In the early stages of the muttonbirding season the young birds are covered in a very light down on top of their feathers. In the second half of the season they come out and rub themselves against ferns and trees and the ground, to remove the down from the feathers, and leave them in their fully fledged state. They're perfectly fit to fly, but until the down is taken off their feathers, they can't.

They rub the down off easily on a wet rainy night. Everything they touch takes off some down. They seem to know this, and come out and go through the process of what we call "tarring". When someone says: "The birds are tarring," you know they're coming out and rubbing this down off on the surrounding bush and ferns.

That's the sign to us to start torching. Sometimes it's in the morning just before the mother bird takes off; at other times they come out in the very early part of the evening, depending on the state of the moon as a rule. The muttonbirder has to work accordingly, morning or night torching.

You take only the young birds. The Maoris are very definite that nothing must happen to the mother bird.

There's never been an efficient machine for plucking the birds. Some of the better pluckers, I've heard, manage 120 an hour. My average is about thirty-five.

In the early days you had to get to these birds while they were warm, because when a bird is plucked hot, any holes left close over. Plucking leaves a bird downy all over. There's a wax we use now that sets like a hard outer shell on the bird, and after a few minutes of setting in the cold air, you take the bird by the neck and the tail, twist, and the wax shell breaks, is easily removed, and leaves the bird perfectly clean.

And then everyone uses salt on the muttonbirds. There's been a big improvement lately in the quality of the salt available, miles ahead of what it used to be. We get really good dry salt that doesn't stick in large blobs when you put it on the muttonbird, so it isn't over-salted now for preserving.

I have heard that muttonbirds migrate to Siberia and to other parts of the world. It's rumoured, though I've never had it proved, that they have been seen in California. However, when I went overseas in 1945, from New Zealand to the Red Sea, every day I saw muttonbirds. I know muttonbirds by sight well enough not to make a mistake. So you needn't ever worry about a shortage of muttonbirds in the shops, as long as the muttonbirders are on the job.

CANTERBURY TALE

R. A. Winn

THE MORNING sun brought scent of gorse, tussock and pine trees up to the rocky top on the Port Hills where I stood balanced against the wind. A noisy plane flew over and shocked the blackbirds into silence; it circled above the sprawl of Christchurch, then dipped to land at the airport.

Haze blurred the city streets but out on the roads beyond, the sun caught an incessant movement of traffic, busy as small mechanical toys. Planes, cars, roads and a city; these are progress and familiar as our own hands, but what was it like, not long ago in time, before there were pinetrees and blackbirds?

Christchurch was a vast swamp, oozy under niggerheads and flax, threaded by rivers and streams and alive with wildfowl. Beyond the swamp were the dry golden plains, with nothing to break the expanse and freedom, and nothing to cast sun shadows in the light-filled air.

At dawn and dusk, wing flutter and bird cries would be the only sounds in an immense silence except when the unhindered winds screamed through the tough matagouri and rattled the strap leaves of cabbage trees.

From the seacoast in the east to the foothills in the west, and a hundred miles from north to south, the Canterbury Plains were formed through the centuries, from greywacke broken by ice and frost and carried down by four great rivers, Waimakariri, Rakaia, Ashburton and Rangitata. Native grasses had never known the chisel edge of animal teeth.

Early settlers from Britain must have longed for the closer horizons they had known at home—and for a more intimate feeling with the land.

Wherever possible, they used rivers as boundaries, and many years were to pass before fences marked possession and plantations broke the distance and took the fierce winds on their backs.

Those first settlers knew the meaning of spadework. They had to dig for water, for sods to build their primitive homes, and for every type of planting. Long distances were walked for stores, and men carried incredible weights before horses and bullocks became means of transport; not all the early settlers had money to bring their own horses.

One man who set up a store in an isolated place was unable to read and write, but he had a method of book-keeping: he made rough drawings of the commodities he sold, if they were not cash transactions. When a debtor called to pay, he knew the bill would be an honest one, but some items took a bit of working out. A puzzled farmer knew he hadn't bought a whole cheese. He and the storekeeper studied the entry, a large circle, and the farmer, after some heavy thinking, remembered that he'd bought a grindstone. The storekeeper had forgotten to draw the handle.

Fencing began with the arrival of wire, the heavy number six wire, so hard to work with that oldtimers instinctively rub the palms of their hands if it's mentioned. In out-of-the-way places there are remnants of it, rusty strands sagging between lichened, grey posts or trailing on the ground.

Sods were cut and built up to form a base for young gorse plants, to mark subdivision, give shelter to stock and, much later, to provide homes to rabbits and headaches to following generations of farmers.

Settlers had to be resourceful and, with-a-saw-and-a-hammer-and-a-gimlet-below, we'll-have-a-new-dray-in-the-morning, was the way they tackled things.

Ceaselessly, shepherds on the large holdings rode through the tussock to tend the flocks, mainly brought from Australia. Chained boundary dogs guarded faraway, strategic points, through hot summers and bitter winters, their desperate loneliness broken by irregular visits from shepherds with old ewe dog-tucker. Their eyes must have strained into the distance and their ears pricked hopefully at any movement that might be the rider they longed to see.

The craze to burn rank tussock, so that young green herbage would spring up in the ashes after the first rain, killed off the native quail and drove a few surviving wekas into the foothills.

Wool was the main commodity and the slow creaking wool waggons carried it, often over trackless ways, to scouring works and wool stores. There were no freezing works, and old sheep were driven over bluffs to their death, or sent to boiling-down plants, the fat making tallow which was exported for cooking and soap-making.

Droving was a hazardous business and the wide snow rivers of the plains took the lives of many men and animals. Sheep had to be forced to cross, and even when headed for the opposite bank, sometimes would turn back in spite of cursing men and willing dogs giving water-logged barks before they were washed down in the current, to scramble out a quarter mile downstream.

Riders learned not to hurry their horses but to trust them to avoid quicksands and pick their cautious way across the river, so deep in places that they had to swim. These were horses of a sturdy breed that could travel over fifty miles a day without any special tucker.

When the rivers grew in volume, and turbulence during nor'westers, and flying grit stung the eyes and hurt the nostrils, sheep would panic and break away. Sometimes it took days to force them across a river.

There were no natural corners where sheep could be held for a night or during a storm; nothing so helpful as the sight that has puzzled many people since, a tall, single-trunk tree with all its branches lopped, except a tuft at the top; a sign for miles, to let drovers know that a holding paddock was available.

Tutu was another risk, and men learned to drive their flocks so fast through infected areas that there wasn't time for stock to graze on the poisonous leaves.

Then came the plough and the great steaming Clydesdales, Shire, and Suffolk Punch horses strained at their chains; down went the tussock and up came the stones. Thousands of loads of stones were picked up by hand and carted away in the old Colonial drays to be dumped in corners. Most of the heaps are there still, hidden by clumps of pine or macrocarpa.

The plains ceased to be formless as fences and plantations made patterns, divided and divided again with the passing years, as the magic word "refrigeration" brought the great breakthrough.

The true sheep-man has been pushed back into the foothills and beyond, and it is he, perhaps, who on a clear day from a place on his high country, can look down on the snug mixed farms, the granary of New Zealand, and see beyond the present into the past because he still stands upon it.

Since time itself is timeless, plainsmen of Canterbury must hold the fabric of the past in strong hands because much of the future could depend on it.

The spirit of it lives in the fine new generation of farmers who, apart from the mechanised age and their scientific agricultural knowledge, retain much of the aptitude of the pioneers in the ability to improvise and invent. They have not lost touch with the soil, and can even take their place among the champion ploughmen of the world.

I watched a railcar streak across the plains, the sun glinting on its windows, and wondered what the men thought who had walked or ridden those endless miles, and whether they had longed for a glimpse of the future as, for a moment, I longed to see the plains as they were then, empty and golden.

As I drove along the foot of the hills, the past vanished as a tractor rattled along at full speed, dogs barking and leaping beside it, and the young driver in shorts singing at the top of his voice the pop song *On the Road Again*. A far cry from *The ploughman homeward plods his weary way.*

MY HIGH COUNTRY FENCING DIPLOMA

Len Moor

BEFORE WE ATTACK this fencing job, let me introduce you to Fred. Fred was a well-educated joker who just couldn't stick town life. Fred took to the backblocks in his late teens and was approaching his half century when I met him. A godsend to any back-country boss, he would tackle any job offering, and make a go of it. Building, blacksmithing, fencing, shearing—a real jack-of-all-trades and master of the lot.

You should have seen his personal kit. A cut-throat razor, made from a piece of steel strap that came around the bales of wool-packs, a shaving brush of goat hair, and tooth brush of pig-bristle, all mounted neatly in finely polished native wood handles.

He had five pieces of different grades of flat riverbed stone, carefully selected in years of searching, for grinding and sharpening. I watched him, one night, slowly punch a hole in an old piece of piano-wire, and patiently hone it down into a beautifully finished sewing needle. What marvellous concentration and perserverance Fred had.

Shortly after I met him, he said to me one day: "I want a mate on a high-country fencing job; how about joining up with me?"

"I haven't a clue about hill fencing," I replied, "but I'd like to have a go at it all the same."

"Just do a bit of fetching and carrying for me, for a day or two, and you'll soon get the hang of it."

It wasn't a new fence we were going to tackle, but one that had gone down in a heavy snowstorm about ten years before, and the boss had just got around to financing its recovery.

There was about three-quarters of a mile of this euchred fence, with two very steep stretches and a reasonably flat bit in the middle, which was the only part accessible to a packhorse; so all the necessary gear would be dumped there, and we'd camp in a hut at the bottom, on the bank of the river.

The first day went in scrambling to the top, and working our way down, assessing the amount of material needed to put the fence back on its feet.

What a flaming mess!

Most of the iron standards were flattened to the ground, or broken off. It passed through patches of bush and fern, where it was completely overgrown, and over small rock faces, where the standards were pushed down into cracks, or holes had to be punched out with hammer and drill.

Fred just scrambled quietly on, making odd notes in his little pocketbook; but my heart grew heavier and heavier as we went on down that ruin of a fence.

Nearly every post was rotted off, and even I, the raw beginner, knew that meant cutting posts from the patches of bush and lugging them on our shoulders, some up that steep top piece for two or three hundred yards—and there'd be a hefty big post for every chain of the fence.

When we got down to camp, Fred couldn't help noticing my long face, and he burst into a cheerful grin. "She'll be a piece of cake, Len," he said, "we'll be on day wages, so we won't have to bust our tripes. I wouldn't have this job on contract for a gold clock; we're set here for three or four weeks. We'll head up to the top in the morning, and put in a few posts until the boss packs in some wire and standards."

I didn't sleep too well that night. I don't mind scrambling around on hills, but lugging broadleaf posts up steep faces didn't seem much like a piece of cake to me. Anyway, we took off next morning with crowbar, shovel and a couple of axes, reaching the top in about three-quarters of an hour.

Fred marked out the spots and set me to digging the holes, while he cut a few posts. You should have seen that chap zig-zagging his way up the steep face, with about eighty pounds of post on his shoulder—a mountain goat had nothing on Fred. He never stopped once and, when he reached me, he wasn't breathing hard enough to blow a fly off a chop.

Out of sheer shame I offered to give him a hand.

"Garn," he said. "I'd rather do this than dig those flaming holes."

"Suits me," I replied. "The holes are right up my alley."

During the day, the boss had packed some of the gear to the middle of the line, so the next morning we were to cart some of it up to the top.

Fred took off with a coil of wire and a bundle of standards; I tried to do the same, but it wasn't long before I was in a state of near-collapse.

"Hey, drop the standards," Fred called back. "We'll just about make it to there with my lot."

Good old Fred; he never slung off at me at any time, and I made some awful blues on that job; but he made a reasonably decent fencer of me. I heard him go crook only once on that job. We were pulling the old wire and standards straight, through a bush section, and they were all overgrown with fern. He made a scathing remark about "This blasted emblem of our flaming country."

Things went fairly well until we came to the one bad gully on the line and here we had to put in a lot of tie-downs, to keep the fence from lifting when strained tight—you've got to strain the wire good and tight in a fence, that's very important. And to make these tie-downs, you just drive an old broken standard in on an angle under the fence, loop a piece of wire in the top hole, and take a half hitch around each wire. It's a sort of anchor.

The top wire was barbed, and strained before being fastened down.

Over the gully, the barbed wire was swinging high, about four feet above the fence at the bottom. It looked pretty odd to me. Had something gone wrong?

Fred explained that we started to tie it down to the fence from each side of the gully, so that if it broke it couldn't fly far uphill and give us a scragging.

We were nearest the road at this point, and this is where Paddy came into the picture. Paddy was a cowman-rouseabout on a farm a bit further up the valley, off on one of his periodic binges and, hearing us above, he kindheartedly climbed up to see if he could do anything in the village for us.

Fred and I were on either side of the gully working to the bottom tying down the barbed wire. Paddy came up the gully bottom, and, before we spotted him, thinking to be helpful, he grabbed the swinging wire, to pull it down to the bottom post. Things certainly happened then!

Ping, it went, and flew like a striking snake, wrapping around him half-a-dozen times, trussing him up like a chicken on a dish.

Fortunately the wire kept low, and didn't get near his face, so he escaped with a few minor scratches, but you should have seen his clothes when we cut him free; they looked as if they had been worried by a litter of month-old pups.

And that wasn't the only trouble we had in that bally gully.

Right at the bottom was a stony creekbed which wouldn't hold a tiedown peg, so Fred fossicked out a fairly big long stone, showed me how to make a sling for it with two pieces of wire, and then tie the fence down to it in the usual way.

Everything was going fine until we strained the rest of the wires tight.

I was just about to strain the last wire, when there was a scraping noise, and a dull clump at the bottom of the gully; then all the wires went slack. What the blazes had happened now?

Off down to the bottom, and there was that dirty great boulder sitting comfortably in the creekbed and my sling swinging in mid-air. I had to put it on cockeyed, lifting one end, letting the sling slide off. Good old Fred didn't go hostile. No, he just grinned, and said, "You're learning all the time, Len."

At last we were on the final strain, and I had visions of a decent-sized pay cheque, which would give my old motorbike a much-needed overhaul, and me a few days amid the bright lights.

This was one of the steepest bits of the line, but nice clear tussock, with all downhill carrying.

Fred had taken off with the post on his shoulder, and I was to bring the rest of the gear.

Clever-boy wasn't going to carry that slippery, cantankerous crowbar. No fear—he had a much better idea. Place the crowbar pointing downhill and nudge it gently along with my boot; it should slide downhill easily. It did. Shouldering the rest of the gear, I gave the bar a push with my toe. Heck, it took off like a lizard! I could hear that blessed crowbar fizzing and hissing, as it shot down that steep face.

Fred jumped sideways, nearly going for a skate with that big post he was carrying on his shoulder. The next thing I heard was a metallic clang, as the crowbar shot over the edge of the bank and hit the riverbed about twenty feet below. Now for a nice little stroll for about half a mile, to where I could get down and then go back and recover my runaway crowbar. When I got there I could see the mark where it had landed in the riverbed, but there was no sign of the bar. I looked up to the top of the bank at Fred—of course, there was that friendly grin.

"Did you find a short cut, and beat me to it?" I raved.

"No, but I'll toss down the shovel, and you can dig it out. It has gone in point first," he told me.

Sure enough, it was a bank of soft shingle, drifted there by the river, and I had to dig down nearly three feet before I got enough grip to drag that blessed crowbar out.

Talk about the last straw breaking the donkey's back. The back wasn't broken . . . no . . . but it certainly felt like a donkey's.

CHAFFCUTTER COOK

Jim Maxwell

LIKE ALL country children I really loved to see the big black traction engine come chuffing down our country roads of South Canterbury. Climbing on to the gate I'd sit watching and waiting to wave to the small chaffcutter train as it trundled by on the narrow gorse-lined gravel road, looking for all the world like a proud mother duck leading her ducklings. It was a thrilling, majestic sight to me, but little did I think that by the time I was eighteen I'd be a cook on a chaffcutter train, touring the countryside and the farms where the beautiful big draught horses lived.

Wherever a lot of horses worked, they needed plenty of chaff, because chaff was the fuel for the big draught horses which were the motive power, the strength of the farms in those days. Good chaff mixed with oats filled the loose boxes of the team in the stable, where the horses would munch and blow before and after their day's work. So a loft or barn full of chaff was a vital necessity in those days.

By the time I was eighteen I got my wish. I did a season as a cook on Artie Johnston's chaffcutting plant at Fairview, just outside Timaru.

Of course I wasn't a proper cook then. No fear—all I had was a first-class Boy Scout knowledge of cooking but I was willing to learn. It wasn't so hard either, as it turned out, with the trusty little black stove tucked away in the corner of the galley, and on the stove stood a huge black iron kettle with a long swan-shaped neck of a spout which whistled and sang to me as it simmered all day.

Work on the chaffcutter came round in the autumn—golden days with a nip in the air and a touch of frost—and in the dull bleak early winter days. We'd leave Fairview early on Monday morning and, weather permitting, wouldn't get back until Friday night. During the week we'd rove far and wide through the countryside of South Canterbury. We'd pull into one farm, set up the big chaffcutter and traction engine, get her going full bore, and work and sweat away till the stacks were finished. Then we'd up and shift camp to the next place, and perhaps work till well after dark so that we could pull out in the morning and start the day somewhere else. I think a stack of straw would

yield about 400 bags, and we could clean up two or three stacks in a day for a cockie.

The chaffcutter itself—green, boxlike, fourteen feet long, with its various attachments such as pulleys, pitchforks, and shovels clipped on to the side—was hitched on behind the traction engine, stoked regularly with good West Coast coal, and puffing out a streamer of smoke as we trundled round from farm to farm.

Attached to the 'cutter was the cook's galley, where two men slept. The galley had a platform in front holding a water-tank, bags of coal and potatoes, and milk cans.

Then finally, fixed at the rear came a Wild Western-like coach like a sleeping caboose where half a dozen men could sleep.

This noisy trundling country train could be heard coming along the road from miles away and, as it passed, the big engine-wheels left their distinctive, diagonal tracks in the dust.

It was a tough but healthy life, and after a week I started to settle in good-oh to the routine. I cooked porridge, chops, eggs, roast mutton, boiled corned "dog" as they called it, spuds, cabbage, carrots, and plenty of hefty swedes from the local farmers' broad paddocks. Puddings: my main puddings were rice, boiled or baked, duffs, and sometimes a ribsticking gooey mass called boiled tapioca, a thing you mercifully never see now.

And for smoko, batches and batches of sour-cream scones which rose fluffy and light. The men, six of 'em, were a hard case gang to cook for. They gave me hell if the food was poor or if it was burnt, or raw or tasteless. Honestly, I really thought they meant it when they threatened to throw me in the nearby creeks!

But between you and me, how I used to love the cosy snugness of the cook's galley on the chaffcutter train. The doorstep and around the stove were well worn and patched with tin, and inside around the stove hung billies, pans, toasting forks, large cooking spoons, dishcloths, and tea towels. At night—*phew,* sometimes they'd hum, too—socks, shirts, hankies, hung over the stove to dry. Two silent beauties smiled at me from calendars on the wall. They never said a word, bless 'em.

There were little cupboards lined with old newspapers for our plates, pannikins, and sugar bowls, and so on, and drawers for knives, forks, and spoons, which slid in and out on well-worn runners. The galley table, white and aged, scrubbed to perfection over the years, had been the scene of many, many card games, drinking bouts, and meals. Along each side of the galley were two forms, worn and polished by hundreds of dungaree trouser-seats. At the end of my galley stood a couple of bunks, or flea-bags (or worse) as the men mostly called them, and they were always littered with a newspaper, magazines, books, or just general clobber.

Besides this galley, there was another little cabin on wheels where the gang slept, but it was a lot smaller. (By the way, there's nothing more comfortable than a good chaff mattress once you get your hip bone in.)

Our foreman was Jerry McKenna. Jerry had worked with Artie Johnston something like twenty-seven years, and wee Teddy Driscoll had been on this machine twenty years. The other three, Norm Vincent, Bill McConkey, and Paddy Sugrue, had been together about five years.

Old Jerry was a fine foreman, and he certainly knew his traction engines.

The grease, oil and grime on his clothes just looked natural for him. At night he lay in his bunk below me smoking "weed", rolling his own and coughing and spluttering his way through the night, passing the time till the welcome dawn came, when he'd get up and go out to his beloved engine to stoke it into life again. And then with oil and grease-gun he'd pamper his machines ready for another full day's work.

Meanwhile I'd be filling the galley with the good old cosy honest smell of fried bacon, chops and taties, toast and tea. The smell would waft through the air to the other galley and draw the men out from under their warm blankets, yawning and stretching.

With a lick and a promise in a bit of cold water in an old bucket handy to the step outside, with a hearty spit and a quick short visit to the nearest gorse hedge, the men were ready for breakfast. It'd be before sunrise: about six o'clock. I'm sure none of 'em ever wore pyjamas, used a comb, or had anything for bootlaces except string from the 'cutter. With a good breakfast under their belts and a cigarette rolled, they were ready to start on the day's work.

Soon there'd be a busy bustling scene around the stack. The farmer would arrive with his horse and dray to cart in the freshly filled chaff sacks, and often one or two kids would climb up on Jerry's traction engine, or come and stare with big round eyes at me working in the galley. Sometimes I'd get them to go and cart buckets of water from the nearby creek for the water tank on the end of my galley. For this I'd give the obliging youngster threepence, or a hot buttered scone which I'd baked for smoko. If it wasn't kids hanging around, it would likely be a curious cow, heifer, horse or pony, or maybe an insinuating hungry dog. And I could always find something to give them to eat, too.

By the time I'd washed up, swept out the floor, scrubbed down that old table, and made a batch of scones or pikelets, it was smoko time already. No nibbling at a dry biscuit: these men could put away three fat scones and a couple of big pannikins of tea, no trouble.

This was hard dusty work, out in the open, buffeted by the cold southerlies or hot westerlies. Chaff, smoke and dust blew and curled around the men's heads, and straw drifted and scattered about the nearby ground. The men loved to curse and banter from stack top to "bag hole" and up to the feeder; there was always plenty of fun and laughter and abuse to make the day go quicker.

As the 'cutter machine grumbled and whined, the belts spun around pulleys, the chaff bags quickly filled, and this kept Teddy and Norm busy threading string through their long bag-needles and sewing up the chaff bags. With a "hip" and a heave they would toss them up on to the great pile behind them.

My job kept me busy, too. Surprising what there was to do, as any housewife will probably agree, but usually I managed to keep up with the demand for food, which came again at twelve and five o'clock.

Sometimes the Master, as our boss Artie Johnston was called by everybody, would stay with us for a meal. He was a big man with a wooden leg, a legacy from the Somme in 1916, and well respected by all the gang.

He'd always say to me: "Feed them well and see they get plenty of tucker." And I did.

Then, by Jove, there'd be some great old-time talk around the table. How I wish I could have recorded all those stories, yarns and memories and laughs about the "good old days on the 'cutter": the time when the traction engine had got stuck in some bog; or when some member of the gang had got a bit too drunk on the local brew of apple cider and had a dose of the DTs.

I finished my season as a 'cutter cook and like the rest of the gang went down the road.

I'll always remember a picture of the 'cutter as it looked one evening when the sky was painted with the pale blues of the South Canterbury winter twilight and the cold snowy mountains ranged along the skyline and the hedges and trees brooded black around the paddocks.

The old Burrel engine chuffed a black column of smoke from its funnel and a fleck of steam escaped from a valve, the big rear wheels shuddered with the whip of the long belt to the 'cutter as old Jerry McKenna stood by for the last bag of chaff to be filled. Then with a long, loud, agonising shriek and hoot of the whistle, he told the world that the chaffcutter had stopped work for ever. The tractor had arrived to take over; and the beautiful big strong draught horses would be needed no more. . . .

Chooks

THE CUNNING CHOOK

Elizabeth Black

OUR MOTLEY COLLECTION of fowls, mainly of the old-fashioned gorgeously-feathered barn-door variety, wandered where they willed, outside the garden—at least in theory. They laid equally gorgeously-coloured yolked eggs, also where they willed, until they died of age or from the undue enthusiasm of a neighbour's dog.

Our old Rover, who looked like an animated hearthrug with four short legs and the woolly head of an old English sheepdog, refused as a rule to acknowledge the existence of the fowls. Sometimes when they came to peck at the remains of his dinner their rapid *tit-tit-titting* on the enamel dish became too much even for his ancient nerves. A throaty rumble would disperse the chooks in a cackling flutter, until the old dog dozed off again.

Then one, then another, would return in that peculiar stalking progress of the domestic fowl which knows perfectly well it is doing something it shouldn't. A one-legged stance, head erect but weaving from side to side, beady little eyes alert; the other leg meantime tucked up, predatory claw bunched, then a quick reversal of legs, and the fowl is just that much nearer to its goal—Rover's plate or Mum's newly-sown vegetable plot; but while in the resting position, a picture of perfect innocence to any suspiciously-minded female human.

THE CHEERFUL CHOOK

John McCaw

KEEPING A FEW FOWLS round the house is one of the pleasures of country living. Everybody used to keep hens at one time, even dwellers in the little cheek-by-jowl cottages that were the pride and joy (and best investment) of the urban developers of pre-war days. Each cottage had only a twenty-foot frontage but a deep section, and most of these sported a small fowlhouse at the rear.

But out in the country we either didn't run to a fowlhouse at all, or we had a ramshackle shed set aside for the birds and they could use it or not as they pleased; there were plenty of trees to roost in. They rambled round the farmyard and nearby fields; they fed on insects and grass seeds, worms, bugs, and horse-droppings. Frequently they got into the house section, picking at the vegetables and dust-bathing in the flowerbeds. But they knew they were trespassing, and fled shrieking at the sight of a hostile woman armed with a carpet broom.

Outside the fence they were in their own country and were afraid of no man, as indeed was only right. For we were friends. We gave them a handful of wheat, even if they had already stolen it. We kept the grain in old cream-cans with battered lids, and often the lids were so

out of shape that they didn't fit snugly and the chooks soon found out that if they perched on them they would knock the lids off.

"As silly as a hen" is a statement often heard, but the statement is as silly as the speaker. Chooks are intelligent creatures, friendly, humorous, and full of deep cunning. Ever tried to find the nest that you were sure existed in the woodheap? You might sit on a stump and watch the heap and the chooks. One would creep into a hole while you were watching, but in a few seconds she would emerge again, far too quickly to have laid an egg, but you would go and have a look just to make sure. There would be no egg, but when you looked round there would be no chook either. She had trotted round the heap and dived into her nesting hide-out when your back was turned.

If you were patient enough she would emerge, quite silently, give you a passing glance, wander off a few feet into some sort of cover, and then burst into song: her egg song—*took-took-took-to-kaw*—repeated twenty times, fortissimo, staccato, and gradually diminuendo.

Should you take her eggs and not leave one in the nest, she would probably go elsewhere and set up another home. You could hoodwink her by leaving a china egg in the nest; but even this bit of easy duping does not prove her stupidity, for I have known of housewives who have attempted to cook a china egg.

There are all sorts of chooks, just as there are all sorts of people; only, generally speaking, chooks are the nicer species. There are some thrusters, gate-crashers—such as the Campine, a virile little speckled fowl with a white neck and head, a sort of Hereford chook. It will squeeze through the tiniest hole in its run or into a garden, and I have seen one climbing up wire-netting like a squirrel. But although Campines are very pretty and good layers, I don't like them: they are too wild and undomesticated.

Most people nowdays prefer the White Leghorn—but I don't. It is a highly productive bird, the magazines tell you, and I believe them too, but I still don't like them. They're not natural. A chook in the Dark Ages was just an ordinary sort of bird which laid a dozen or so eggs and then sat on them to bring out a clutch of chickens. But scientific poultrymen got hold of the White Leghorn, turned her into an egg-laying machine and clapped her in irons if she went clucky, so that she has lost all her maternal instincts but goes on laying eggs till she looks like a moth-eaten caricature of a hen.

I know one Kiwi farmer who does like them, though. He'd been on a world tour, and when he came back to his farmyard and saw his mangy half-moulted White Leghorns foraging about, he muttered: "You'll do me. Everywhere I went in Israel, I saw your half-sisters in all the settlements, and what's good enough for those bright boys in Israel is good enough for a Kiwi."

But give *me* a big Silver Wyandotte, or a great Rhode Island Red, or a heavy Black Orpington; for here you have a chook, a real solid sturdy citizen, peaceful, contented, happy. They are always singing a cheerful little aria. They run up to you and crowd round your feet, will even climb on to your knees and shoulders if you encourage them. They will talk to you—impatiently if you are slow with their food, cheerfully if you are quick with it, contentedly when it's all eaten, and sagely if you engage them in conversation.

I have known several people who told the daily news to their chooks, just as many beekeepers tell their bees the happenings of the day. One old lady assured me that during the First World War she found that her fowls responded to good news by laying more freely than when the war news was bad. I suppose her hens laid two eggs each on Armistice Day.

The heavy breeds nearly all lay brown eggs, and who wouldn't prefer a warm-looking brown egg to a pale anaemic white one? The farmyard hen laid an egg rich in vitamins and with a deeply-coloured yolk, so unlike the watery-looking feeble yolks of the battery hen. They had real health and vigour, those eggs.

A very corpulent priest of my acquaintance met one of his parishioners in the street, a dear lady who was solicitous of his health. "Indeed, Bridget," he replied, "I've no' been well at all, at all. I could manage only one egg for breakfast today."

"Oh, dearie me," said the old lady. "And what good would one egg be stradagin' about in a great creatur' like you?"

But, you see, it was a farmyard egg, probably a brown one, and it kept the big fellow going all day.

I have been given a brightly illustrated brochure, turned out by the

Tourist Department of the Argentine Government for English visitors. It has been written by a fellow who evidently thought his English was pretty good. Here is how he describes the peasant farmer, his beliefs and customs, and his poultry:

> Those who their abodes in the country, are always aware of the mysteries surrounding them, are always on the lookout for them; they are afraid of the occult forces and their ominous influences, and divine their impending nearness through the signs that nature everywhere seems to be giving them. Many of these omens they consider as signs of some impending misfortune, and among those there are some they attribute to the hen, the most useful among the domestic fowl. If the hen cackles like a rooster, a misfortune is sure to be near. If the rooster chants three times before sunset, or if he chants an odd number of times between eight and ten o'clock of the evening, this is also a bad sign. Then the face of the farmer will become somber and an expectant fear will be troubling him.

But back to New Zealand and my barnyard fowls—with no expectant fears to trouble us.

I mentioned that chooks have a sense of humour. Have you ever tripped and fallen into a pan of household scraps and had the hens gather round you, after their initial fright, to gaze at you with heads on one side, the glint of laughter in their beady eyes? Or seen them with a weta? They stand off in horror, but so fascinated they cannot leave it. They close their ranks until shoulder to shoulder they form a ring round the creature—and then the fun starts. None of the hens is brave enough to attack the weta, but they all try to encourage their neighbour by pushing her forward, trying at the same time to keep themselves out of the firing line and with a clear route of retreat; all the time dancing about with palpable pleasure.

Yes, keeping a few chooks round the house can be great fun. But keep them round, or outside, the house—not in it.

I have visited plenty of places where the fowls lived under the house or on the verandah or even in a disused store-room. But one homestead I knew had been taken over by hundreds of hens kept by the owners.

I saw dozens of hens and ducks in the house, walking through it with muddy feet, perching on tables, beds, sideboards and mantelpieces. And the marks of their occupancy were obvious.

And now, please excuse me. I'm going outside. Guess where? Of course—outside to read this to my own few chooks, before popping it in the mail to "Open Country"—that is, if the chooks approve, of course.

THE CAPTIVATING CHOOK

Enga Washbourn

A HEN WITH a personality that lifts her out of the usual run of scuffling, scatty, rattleheaded, suspicious, malodorous, but indispensable birds, is rare indeed. The odd few outstanding characters that we've been privileged to know could be counted on the fingers of one claw, but they remain stored in our memories with affection.

Blackie was a large fluffy Australorp, reared with five sisters from day-old chicks: she was the only one of the family to emerge with character and originality. At that time we were living beside the beach in Collingwood at the top of the South Island, where my father was the doctor during the war. In due course these chicks grew up, six fine upstanding birds with iridescent green-and-black plumage, shining brown eyes, and healthy red combs. Nests were prepared, and we began to watch eagerly for the first egg. It was at this stage Blackie decided not to be an ordinary hen: the nesting-box was not for her.

While her sisters retired dutifully each day to lay a large brown egg in a box of clean hay, Blackie went about the business in quite another way. Sharp at 6.30 each morning she strutted purposefully across the road from the henhouse beside the beach, up the steps and along the verandah to where our brother, home on leave from bloody battles at sea, was sleeping peacefully in a partially-enclosed verandah bedroom. Leaping lightly with a flutter of feathers, she landed on his feet, and from this launching pad took off into a wardrobe on the far side of the bed. Here, in concentrated silence, she deposited a warm pinky-brown egg on a pile of clothes on a shelf, and departed cackling with pleasure. An hour later this fine fresh egg appeared with bacon on John's plate for breakfast; and far be it from us to discourage Blackie in any way.

When at the end of his leave John left for a destination unknown, Blackie also left off laying in the wardrobe.

Her next choice of nests was one of the chairs in the surgery waiting-room. By then she was laying later in the morning, and it so happened that this coincided with the consulting hour. Quite unmoved by doorbells and the arrival of patients, at 11 o'clock each morning, come hell or high water, Blackie strode determinedly up the steps into the waiting

room and fluttered on to her chosen seat. Surrounded by slightly startled patients she laid her egg, neatly and efficiently and without any fuss, then went off down the steps chortling to herself to scratch for sand-hoppers on the wet sand. We tried, when possible, to warn newcomers to the surgery what to expect—more for Blackie's sake than the patient's, as we didn't want her disturbed. The regular patients took this daily spectacle in their stride with tolerant amusement.

The next character of interest to join the family emerged from a setting of eggs of mixed barnyard origin.

From one egg came a bird of great charm and gentleness, more like a weka to look at than an ordinary domestic hen. Not only did she resemble a weka in appearance, but she acted like one: so much so that in due course we consulted an expert to ask if a cross between a hen and a weka were possible. We were disappointed to be told that it was most unlikely. However, be that as it may, we always called her Wecky, and for the few years of her short life she was carried around, cuddled and petted, even accompanying the children on to the beach for walks, or sunbathing, when she lay stretched out beside them.

She had no idea that she was a bird, regarding herself as one of the family, and when at an early age she died of some mysterious disease after a short but harrowing illness, she was buried amidst floods of tears in a peaceful spot overlooking the lake.

Three Austral-whites of the same vintage as Wecky belonged to the ten-year-old daughter of the house. She had named them after her three current favourite authors—Jane Austen, Elizabeth Goudge, and Enid Blyton. Of these three hens, Enid had easily the most character, but not all of it good, by any means.

She was a stocky bird with rather short legs and a pretty design of black spots scattered over her white feathers. She had a friendly, placid nature, and she knew which side her bread was buttered. She grew up pampered, fondled, photographed from every angle, and overfed by her doting mistress. After laying a few eggs she became broody, so a peaceful nest of hay was made for her in a dark corner of the car shed.

At last the hatching day dawned. Correspondence lessons were taken into the car shed, and Enid's young owner curled up beside the nest to watch proceedings, poking crumbs of iced sponge cake into her beak at frequent intervals. Finally the shells began to crack, and out climbed the cheeping, damp little chickens.

At this stage Enid lost her head completely and blotted her copybook for all time. Far from displaying any motherly instincts towards these helpless little balls of fluff, she began pecking them fiercely, and even attempted to eat them—to the enraged disillusionment of the watching owner, who rushed in for help.

Enid's end was colourful and unusual, too. She was attacked by a sparrow hawk and carried high up the hillside behind the house, where weeks later the pitiful feathery remains were discovered and her sudden and mysterious disappearance was explained.

A friend visiting us one day put our own feelings into an eggshell when talking of his own hens. We asked rather uncertainly whether he ever *ate* his own birds.

He looked at us with horror: "Eat our own birds? Why, we can hardly bear to eat our own *eggs*!"

THE COLOSSAL CHOOK

Edmund Lane

Edmund Lane of Opua, in the Bay of Islands, is sitting at his breakfast table. Between his hands he holds an egg. It's easily as big as your two fists put together and it has speckled brown spots all over it. Mr Lane is looking at some writing on this superb egg:

IT'S MY MOTHER's handwriting, and it reads: "Judy's first egg, December the 2nd 1904."

And who's this wonderful Judy?

She was an ostrich. We had ten ostriches. In those days ostrich feathers were all the fashion—ostrich feathers in women's hats and feather boas round their necks. We had wonderful success in our curling feathers and feather boas—we got as much as 17s 6d for a good curled feather, and £5 for a feather boa.

What exactly is a boa—a sort of super fan of feathers?

Well now, it's all the small feathers teased and wrapped round a cord to make . . . . er . . . . what do they call them—to go round your neck.

Oh . . . . one of those—what the blazes is it—I don't know—like a sort of super-scarf?

Yes, a sort of scarf, a feather scarf that goes round your neck. It used to be all the rage in 1900, and around that time. Anyhow, we had ten of these ostriches, and they all had to be boxed securely before we could get their feathers off.

Before we get on the shearing, plucking, or mowing of our fine ostriches, how did they come up to Opua? How did you come to get them, and transport them?

They came up by ship, the old *Clansman,* in horse-boxes. We had barges and a launch waiting to ferry them ashore, still in these horse-boxes, and they were taken over to the farm and let loose into the paddock. The horse-boxes were returned to the ship.

Those ostriches didn't come from the zoo at Auckland?

No, no. They came from L. D. Nathan, from Whitford Park in Auckland. You see, my father had a 2,000 acre Crown lease of land,

and he was busy clearing this farm. When we visited Auckland, they told him that ostriches would be a great asset in clearing the farm—they would eat anything. Well, that was quite true. They *would* eat anything; but they didn't like teatree and fern very much, but they did like soap and nails—anything like that—anything that glittered, and tennis balls too: they'd chase tennis balls, collect them up, and you'd see the ball going down their necks.

In 1904 we got the ostriches, and we had them for about six years. They cost a hundred guineas a pair. We reared many chickens, and we sold quite a few chicks for ten guineas each, but that didn't compensate for the losses we had.

They were the most silly things imaginable as far as weather and weather conditions were concerned. In the paddock where they ran were several old kumara pits which just fitted their bodies nicely. If it rained, that was just too bad: they'd fill up with rain and these ostriches would be stranded, quite unable to get up for cramp.

Quite stupid?

Absolutely stupid. The paddock itself was railed right round with

teatree rails. After a few years the ends of these rails where they were bound together would curl up. The ostrich always felt there was better grass outside the paddock than there was in it, and several of them got their necks caught in these joints and hung themselves.

Did you give them much food—mash or grit?

Yes, we had to feed them. At night they were brought into a smaller paddock where they were housed, alongside the homestead. But in spite of our care, when they were let out during the daytime they would do these silly things.

You didn't sell any eggs or sell any of the meat? As poultry?

No. I wouldn't care to eat ostrich very much—I don't think they'd be marketable.

Can you remember the names of some of them?

We had Jack and Jill and Judy and Peter. I think we had about three male birds. All the rest were females. They all had names.

And you'd take these great plumes from the tails; they could look very beautiful I imagine?

Yes, the best plumes came from the tail. Several good plumes came out of the wings, but the tail plumes were the ones that were fancied. They used to be curled, and those were the ones that you got the high prices for.

I've often wondered if the ostrich looked a bit dejected after you'd taken his finery off him.

Well, I don't know whether you could tell much by his looks, but I've no doubt he had feelings.

How did you pluck them, removing these feathers? And were all the feathers from the tail? I believe they can kick like billy-oh?

Yes they have a traverse of about eight feet in their kicking, so the only way to safeguard yourself is to have a forked stick about ten or twelve feet long, and if they came at you, get their head caught in this forked stick, and you've got them.

Have you got kicked yourself?

No I haven't but my mother got kicked very badly on the head. She never got over it.

Very powerful?

Here's an example of how powerful they are. One day my father was riding through the paddock on a light draughthorse when this male bird came along and caught the horse on the jaw, and tore the jaw out. The horse had to be destroyed. But when father went to get up, he found the ostrich standing over him. How on earth could he get out of this? Anyway he grabbed one of the ostrich's legs. The ostrich in the struggling fell over, so immediately my dad jumped up and ran for the nearest cliff. He only just got over the cliff when the ostrich appeared above him. That's how quick they are. An ostrich has been checked at sixty miles an hour. They're terrifically fast.

How would you go about plucking them then? You'd have to be careful.

Too true! You had to build a box to hold their body. You drive them into this box, shut the door, and put a hood over their head because they can peck. When you're plucking them you have to use a form of secateurs to cut the feathers, because pulling them out of the sockets would be too painful, and upset the birds. As for the tail feathers—that's where the beautiful white ones come from—one or two are pulled, the rest are cut.

Would you shear or pluck once or twice a year?

Once a year. It's quite a business when you've got ten of them to do: quite an art in getting them into this box, using this long forked stick and guiding them in with this forked stick round their necks. And you've got to have a small sort of holding-paddock to confine them a bit before you can put them in this box.

Did they have any special feed? And did they call out at night or make any strange noises?

No, not really—there wasn't a great deal of noise connected with them.

Did they crow or cackle or go moulting?

They did have a sort of a roar—like a lion. That was the male bird. One time, just across the inlet, there was a road that used to go from Russell to Waikare. The Maoris used this quite a lot in the evenings going to Russell—of course that's where the pub was. One night the male bird started to roar when two or three Maoris were going along this road just opposite the farm. They thought the taipo was after them, and they never stopped galloping until they got back into Russell.

Why did it all finish? I hope it had a happy ending in a way, Mr Lane?

It didn't really. They were such silly things. The mortality rate was very high, the bottom dropped out of prices when the ladies' fashions changed, and we finished up by selling the last four or five ostriches which survived. We went out of ostriches then, and decided to go in for cattle and sheep. It was quite a relief!

Up In The Clouds, Down In The Mud

HIGH OPEN COUNTRY

R. A. Winn

THE RED-BROWN ROCKS glowed through the snow as the last of the sun's light touched the peak of Malte Brun, over ten thousand feet, in the pink-streaked sky.

Shadows darkened below and, far down, the long Tasman Glacier turned blue and cold.

From inside the hut one of the party called: "Tucker's ready! What's it look like for tomorrow?"

"Not a nor'west cloud in sight," I said. "It should be all set."

I shut the hut door against the evening chill, and we ate and smoked by the light of a candle held by three nails in a block of wood.

At 4 am the old alarm clock clanged around the iron walls and brought us all out on to the floor.

As we left the hut the air flowed past us in waves of cold, but we soon warmed up to the rhythmic crunch of crampons as we climbed westwards towards the Divide. Even with friends on a mountain under the morning star there is a solitude of soul and a strange quiet that makes the world remote. It was only with the coming of the sun that we emerged from our private selves and became a team.

We were two days out from the Hermitage, now about eighteen miles behind us to the south.

Two thousand feet above, the skyline gave no indication of the contrast beyond when we'd look down on the shaded greens of bush-clad Westland. We stopped to look south and east over the rock and tussock of the great Mackenzie Basin.

Trees marked the nearer sheep stations, but the faraway ones were lost in the haze of distance. Our round trip would take us about forty miles on a course west, then north, and finally south, to land us back at the Hermitage.

After four hours of slogging on the Haast Glacier and crossing many deep crevasses, where we felt the comfort of the rope, we reached a point just below Pioneer Pass where a crevasse big enough to drop a house in blocked our route; but we found another way and reached the Divide at Governor's Col.

On Grey Peak we needed a spell. We dropped our packs. While we ate cold chops we looked west to the Tasman Sea, glassy and flat from our mountain top. There was no sound nor any movement on that still morning. A small, solitary cloud hung motionless in the pale sky, and a dead butterfly, like a wisp of black paper, lay on the snow, blown by an upward wind far above its place of happy living. It was a mountain butterfly, *Pluto,* black to attract and hold heat.

The long plod over the soft snow on the Fox Glacier was hard going in the mountain heat, and the old Pioneer Hut was a welcome sight. This hut, subsequently swept away by a fall of rock, was perched on a narrow man-made platform at about eight and a half thousand feet. The spur fell away steeply on two sides, and an unstable rock ridge towered behind to the skyline. To get through the low doorway we had to shed our packs and crouch down.

I'd heard tales about this refuge, and they weren't exaggerated; they couldn't be. Along one wall a cupboard, a bench and a kerosene cooker crowded together, with about two feet of floor space to the door. The rest of the hut was taken up by a platform, known as a "Maori bed", to hold four men.

There were four in our party. I looked closely at the bed to select my quarter share; caution made me look to the ceiling where I saw broken rafters roughly spliced and a new sheet of iron on the roof.

Only one thing could have made that hole, and it wasn't mice; so I chose a spot as far as possible from the line of fire and dumped my sleeping bag on it. The other two did the same before the guide came in. He gave us a significant stare, then looked at the ceiling and the vacant space on the bed.

"I might have known you'd leave that space for me; a boulder as big as a bag of sugar fell through the roof last time I was here."

We all laughed, and he grinned rather ruefully.

We had a quick meal and prepared for an early night. Buckets of snow were brought in for a morning boil-up, and we crawled into our sleeping bags. Before the candle was blown out I read in the hut book about parties who had sat out days and nights of fierce storms, when only the double strands of fencing wire over the roof had prevented the hut from being blown away.

We were lucky that it was a quiet night, and we woke to a perfect morning.

Crampons were strapped on at the door and we set off across the Fox Snowfield for Marcel Col. We reached it in bright sunshine and climbed an ice-ridge to a point where the three peaks of Haast stand out into Westland from the Main Divide.

The take-off was a narrow snow ridge which we straddled to reach

sun-warmed rocks, and we climbed along them to the high peak. At 10,295 feet we were high up in New Zealand but could still look higher to the pure sharp peaks of Cook and Tasman. Avalanches thundering down thousands of feet to the Grand Plateau provided the perfect orchestration for the dramatic scene.

We unroped and, with our heads resting on our packs, stretched full length on the rock slabs which formed the few feet of the peak top. Mentally and physically we knew a drowsy satisfaction as we gave ourselves up to the peace of the hour. I wondered sleepily if Heaven felt like this and whether any climber had ever had an evil thought on a mountain peak.

I was almost asleep when: "Time to get cracking!" shook me back to the moment.

Some photographs were taken and then we roped up and started the descent. The crisp snow surface of morning had turned slushy under the sun, and it seemed a long way back to the Pioneer Hut. We were glad to get inside out of the glare and, over mugs of tea, to live again the details of the climb. It had registered with each of us in a different way. I think it was Mark Twain who said that there is no bad weather, but only different sorts of good weather. We'd struck the sort that suited us and decided to make the most of it and climb Glacier Peak next day.

That meant another early night, so we turned in at dusk. Some time later our guide found it necessary to go outside. There were no mod cons in the Pioneer Hut, and the only place to go was over the edge of the built-up platform, on to a ledge jutting out thirty feet below. He stuck his feet as far as they'd go into an ancient pair of tennis shoes a size too small, and with laces trailing and a candle in his hand he went out into the night.

A quarter of an hour later I wondered why he hadn't returned, so I went to the door to look for him.

The beam from my torch caught his face as it came up over the parapet.

"You had me worried," I said.

"What about *my* worry?" he grunted. "I've just had the climb of a ruddy lifetime and thought I'd have to call for help."

We heard his story.

When we'd arrived at the hut, our youngest member had heated the frying-pan to clean out stale fat and, unknown to us, had tipped the slippery mess down the rock wall, where it had congealed. The guide had tried to climb back—in smooth tennis shoes—and it became one step up and two down. At the first slip he'd lost the candle. We went to sleep laughing, in pure mountain air faintly tainted with rancid grease.

We had two more days in weather that climbers dream about, and began our homeward journey by crossing to the head of the Franz Josef Glacier and on to Graham's Saddle, then into a blizzard on the Canterbury side. The cold was intense, and ice formed on our hair and made the rope awkward to handle. Broken snow and ice forced us to take the rock ridge leading down from De la Bêche, and it took a long time to reach the comparative calm of the Rudolph Glacier below.

We had about eight miles to go down the Tasman Glacier to reach the Ball Hut. Halfway down we saw the Hermitage bus, so we took the pack from the youngest man, who hurried ahead to ask the driver to wait for us.

We were very tired, and longed for a lift and a hot bath. Wearing crampons for hours at a time, especially on steep slopes, puts a heavy strain on ankles and calf muscles, and it was with relief that we reached the comparative level of the Tasman Glacier.

Our runner was ten minutes too late to hold the bus, so we went into the Ball Hut for the night. We were out of food and were resigned to a hungry night, when we spotted an old-fashioned Doulton basin covered by a tea towel.

Inside was lemon jelly, quarts of it, set and ready to eat. Unless it's made with the fruit that interests and the wine that soothes, jelly doesn't appeal to me, but this lot made me understand Oliver Twist. We gathered round, weary elbows on the table, and got busy with tablespoons. Nobody spoke while we mopped up the lot and even scraped out the gritty bits at the bottom. We didn't talk that night, but sank into deep sleep beside the empty bowl.

Early next morning we rang the Hermitage, after tinkering with a reluctant telephone, and asked if a bus could be sent for us. Post-war petrol rationing was on, and the refusal was polite but firm. We tightened our stomach muscles and marched the thirteen miles to the Hermitage and drank reviving lager before lunch.

The round trip, two crossings of the main Divide and three peaks, had become a shared memory.

Next morning, as we loaded our packs on the bus for Christchurch, the guide came over to say goodbye.

"Just as well you chaps are leaving today, or you'd be as unpopular as I am. That basin of jelly we cleaned up was meant to be the second course for seventeen tourists at lunch today."

His eyes twinkled and didn't express regret; neither did I feel it. Our need was greater than theirs; and since then I've had a genuine respect for jelly as a life-saving food.

THE WEDDING DANCE

Asquith Thomson

WE'D BEEN GIVEN rough directions how to find the hall, and in daylight a few wrong turns wouldn't have mattered much. But now we'd been following our headlights for twenty minutes without getting anywhere; so in the mild darkness of the autumn evening I stopped the car at a crossroads.

"What are we stopping for?" asked my wife. "They told us we'd go over a stone bridge. We cross a stone bridge and turn right."

"There might be a signpost," I said as I wound down the window. But there was only a willow tree reaching over a gorse hedge. It was very quiet. For a wee while we just sat there. The green clock on the dashboard ticked away like a mouse running across paper. Then suddenly the car was filled with light as another came up behind us; a big, dusty, powerful thing.

As it came alongside it slowed down, and a friendly but unfamiliar voice called out: "Jan's wedding dance?"

"Yes," I called back. "Yes."

"Follow us," said the voice, and the car was past and accelerating.

I let in the clutch, thankful for being rescued from a web of anonymous roads, and followed. We found the stone bridge we'd missed earlier on, turned right as they said we would in a wide curve, then past the grocer's, the school and the post office, and there in a pool of light was the hall where the dance was. On the gravelled space around it some sixty or seventy cars were parked, country fashion, just where their drivers had stopped. It was a traffic officer's nightmare, and no-one gave it a single thought.

Inside the hall was brightly lit; none of that city dance-hall nonsense where pale townies wear dark glasses to keep out the glare of a single candle in a greasy bottle. There was none of that. Everyone could see who he was talking to; in fact, in any corner you might have read the small print of a Customs Declaration. The hall smelled of badminton, A & P home exhibits, Plunket Day meetings, and French chalk. I looked fearfully towards the band, but it was all right: there was none of the deafening clutter of cables, amplifiers and microphones beloved by city bands; we were given the credit for normal hearing. Piano, saxophone and drums were prepared to stand or fall by their own human efforts.

The hall was already well filled, and the guests who were sitting on chairs and forms around three sides of it were already at that stage of free and happy conversation that is reached at a town party only when it's almost time to go home. Yet—and here's something I noticed through the night—no one had more than just something to drink. Many had nothing at all. The pleasure, good humour and excitement of the occasion were enough.

One or two dances went by. Then, over the general hubbub the MC called: "Take your partners for the Alberts, please."

There was a burst of clapping and everyone rushed to form their sets.

"With us, with us," some cried as they tugged an uncertain couple into their group.

A young girl screamed happily as her partner swung her off her feet in a few practice spins. People were everywhere, crisscrossing, sidestepping, and threading through. The floor was a veritable cat's-cradle of dancers.

My wife touched my sleeve. "What do you think?" she asked.

"Us?" I said. "No, no—I've forgotten the figures. I'm a Destiny man, myself. Let's wait for the Destiny."

"One more couple down there," indicated the MC; and since by then we were half-sitting, half-rising, he added, "Thank you. Gentleman with the lady in the blue dress. Over there by the window. Thank you." And obediently we took our place; on the side, fortunately.

There was a little more rearrangement, then the MC called, "Are you all ready?"

"Yes," roared the dancers.

"Then let 'er go," shouted the MC. "Honour your partners!"

And we did that. We all bowed while the band played a long chord, then away we went.

"First and second couples: half right, half left, swi-i-ing your partners and ladies chain! First gent's solo. Swing corners and promenade. All ladies to the centre. Gents hands across and swing!"

This was really dancing. You took hold of people, of hands and arms and shoulders and waists, and the liveliness inside everyone you touched flowed through you. Better than wine. If sometimes I got lost a bit, it was only for a moment; someone always put me right, and on and on we went, with hooches and whistles and stamping, till at last my wife and I were together again, simmering down in the circular waltz.

"How are you going?" she said.

"Bearing up," said I, "bearing up magnificently. Fred Astaire had better watch out."

When the next dance was announced, I happened to be talking to a beaming Mum in a bib apron who had appeared before me with a great jug of cordial.

"You're from town," she declared as she poured out the lemon-floating drink.

"Yes," I said, "but still friends of the bride and groom. My name's Thomson."

"Mine's Walker," she replied, and with a toss of her head but not looking round she added, "that's Dad down by the door. We've been in the district twenty-six years. Came up from Wyndham."

"I'd like my wife to meet you," I said, turning to the chair behind me. But the chair was empty. "She's gone," I said, "I don't know. . . ."

"I know," said Mrs Walker with a broad wink. "She's dancing with young Keith Mitchell."

And, by heavens, so she was. Gone! From right under my nose. This was lovely! Obviously, in Keakea you didn't dance with the same partner all night. So, if it was good enough for Margaret. . . .

I looked about me.

I'd have liked to dance with a young dark girl who reminded me of someone I'd once known; but at the last minute my courage failed me, and instead I asked the woman sitting alongside her. She might have been the girl's mother.

But no matter. We got on fine together, and when the band played *Me and My Shadow* we both sang the words, though at the same time I was also keeping an eye on young Keith Mitchell. He was talking to my wife like a politician in a marginal seat.

Well, since Margaret had no dearth of partners, I skipped a few dances after that and wandered about for a bit.

Kids! There were kids everywhere, all having a whale of a time. Unlike city people attending a dance or a party (and this was a combination of both), it never occurred to these country folk to leave their children behind, in charge of a bored babysitter at thirty cents an hour and all the heaters on FULL. Not on your life; the kids came too. Even the babies, who lay snug in prams and carry-cots in a room off the kitchen. If any of them wakened, then one of the many Mrs Walkers joggled and shushed them till they fell asleep again. Occasionally one of the real mothers was called in, though how anyone knew which mother to call for which baby was quite beyond me.

In another anteroom a horde of young children leaped and lunged in fine, noisy, bewildering games of table-tennis, and so far as I could make out all the kids older than the ping-pong players took part in the general dancing. Not every dance, of course, but pretty often. Fathers waltzed with schoolgirl daughters eager to be grown up. Mothers manoeuvred their shy sons about the floor, even though in theory the sons did the leading. Some children danced with their grandparents. Everybody danced; even the hall itself, as the streamers swayed overhead and the floor bounced under the dancers' feet.

Up until now, and it was getting towards ten, the bride and bridegroom had not appeared. Then, after a longer than usual pause between dances, they did: the bride in her satin dress, though without the veil, on the arm of her smiling, husky, six-foot husband. Straight away they were set upon with smiles and handshakes and slaps on the back, and wise and foolish advice, and hugs and kisses. Then, having been topdressed with confetti, the bridal couple led off the Grand March, and after we'd waltzed that to a finish supper was served.

In the hall? No, no. In the supper room? No, no, no. Outside, out of doors, in the black still night, under a faint gleam of scattered stars. For, while we'd been Grand Marching, trestle tables had been set out and everything we could wish to eat at that hour, and more, was set upon them.

We ate standing up and walking about. We ate roast chicken, and bacon-and-egg pies, and salmon and oyster patties, while talking to solid farmers about the benefits of burning off wheat stubble. Chatting with total strangers, to discover we had close mutual friends, we ate small potatoes roasted in their jackets and with butter like golden lava spilling down their cracked sides. Smiling girls materialised out of the dark with trays of savouries or chocolate cake or wedding cake. We mingled with groups of young people who, in between stealing private kisses—for which they were always publicly applauded—consumed large helpings of fruit salad and pavlova.

All around us the tops were fizzing off soft-drink bottles, but, true to our years, my wife and I settled for tea, which many Mrs Walkers were pouring out at just the right temperature for drinking. Then gradually, in twos and threes and happy groups we drifted back inside the hall.

Relays of shirt sleeved and ruddy young men were just finishing repolishing the floor. They had no polishers—not even a broom. Instead they used strips of broad sacking weighted with clusters of children, who clung to each other like kittens in a basket as they were dragged up and down. When the job was done, the laughing kids were swung off their magic carpets by centrifugal force and the dancing began all over again.

Sometimes in the course of this figure or that I found myself with Margaret for a partner, but for the most part I danced with other men's wives and, when I could manage it without being too obvious about it, with their pretty daughters. Some people might claim that these country girls were old fashioned. They were, I suppose, if by old fashioned you mean that they didn't have kohl black eyes, or green eyelids, or flourwhite faces drained of any expression . . . but I was thankful for it. Not one of them was brittle, or sulky, or afraid to smile, and some even went so far as to laugh outright at my fatherly jokes.

Oh, dear! The years fell away from me and I almost convinced myself I was only twenty-two.

But if my heart imagined I was twenty-two, my legs didn't, and by one o'clock I was weary. So was Margaret. We sat out for a while, hoping to recover, but instead of regaining any strength the little we had ebbed away. We'd had enough. So, after saying "goodnight" to a great many more people than we'd said "good evening" to, we found the car and drove back to the city at an easy pace.

Over the stone bridge we went, and down through the dark still pine plantation that was now fast asleep, neither of us saying very much. An occasional rabbit lolloped along in the gleam of our headlights. But our not speaking was an amiable silence, for we were still on top of the world. We were just bone-weary. The pleasant re-talking of the night could wait.

You'd like to know where and how long ago that wedding dance was? Only two years ago, here in South Canterbury.

THE SWAMPMEN OF HAURAKI

Walter Christie

WHEN WE READ in the newspapers that the Government of the time (I think it was the Massey administration) had decided to reclaim a vast area of peat-saturated swamp, the sceptics shook their heads and predicted disaster for the country if the project was carried out.

But the engineers of this immense undertaking must have been very farsighted and clever men. To the average person the mere thought of the work ahead was appalling.

Go back with me in imagination and stand or sit on one the many surrounding hilltops, and feast your eyes on an expanse of bog and marshland that, wherever you looked, disappeared into the blue haze of the distance, even if you used fieldglasses. If you gazed towards the south-west, a vast morass of flax and raupo swamp appeared, and through this three rivers flowed on their sluggish way into the Hauraki Gulf. The waters of two of these rivers seemed to be hardly moving.

So just think of a project like the reclamation of what is now known as Hauraki Plains in Auckland Province. Imagine the hundreds of miles of drains needed to dry out this enormous area of bogs and water. And remember, every foot of this work had to be laboured by man with a puny shovel; and a shovel only moves a square foot at a time. No wonder this great undertaking took years to complete.

But it was completed, with just shovels and barrows, though many heads were shaken and many predicted failure and bankruptcy.

Even to drain your own quarter-acre section you'd get better equipment than these men had for the Hauraki Plains scheme. But in spite of floods and storms they carried it through to completion. Nothing can stop men of determination and imagination.

To drain the water off the land, contracts were let to Indian and Yugoslav immigrants, in my opinion the finest drainers in the world. I have worked alongside them in digging and cleaning work, and always found them honest workers. Perhaps a little more so than some of my colonial brothers.

I was working on these Hauraki drains in 1920 at the Morrinsville end of this huge undertaking. My companions were two Dalmatians, and we worked together for two years. We walked night and morning four miles to work over the swamp, wearing ordinary leather boots. During working hours, however, working in the drains we wore thigh boots—when we could afford them: those thigh boots cost thirty-five shillings a pair then. At the end of a working day we would leave the boots stuck on top of tea-tree sticks shoved in the side of the drain bank.

We bached and cooked our own food, mostly boiled mutton for all of us drain-diggers. Occasionally if we were off work on a Sunday we would have a roast. But during the week we would have toast for breakfast; the mid-day meal which we took to work with us was a four-pound loaf and a couple of pounds of cheese each. A man would eat this without any trouble; the running water seemed to give you a tremendous appetite.

We did our own washing in a small stream close by our shack, and hung it on nearby willow trees to dry. Some of the men on this job were a lot worse off than we were: they lived in tents.

The only tools used on this work were sharpened round-mouthed shovels, short-handled shovels with the top cornice of each blade cut off at an angle. This was done so that the shovel could be withdrawn easily from the masses of fibrous roots which made up these peat swamps. Pretty primitive, eh? Each block of peat was dug out by a three-action cut, sides and back; every piece would be as big as a butter box and had to be thrown up and out, in such a way that it landed six feet away from the edge of the drain, so as not to press the banks in. This required hard manual labour, and a certain amount of skill, I can tell you.

After a day's work of this kind I can well remember that at times we were so tired and so exhausted from the digging that we used to toss up whether we would go home to our shack or sleep on the bank of the drain.

There were hundreds of parties of workers on these drainage schemes. A good many of them you would never meet or see. There was no social life because most of us had had enough by the end of the day, and after a meal would turn thankfully in to bed. Sometimes I would think: "What a fool I am to be stuck into this backbreaking work."

It was brutally hard labour, but I never gave in because at this stage of my life I was in the process of forgetting a sad bereavement, and this work was the best remedy.

During the course of the day my mates and I would joke and jest with one another; for instance, if one of us disappeared into a hole at the bottom of a drain where a stump had just been blown out. All the bottom of the drain was sloppy mud, and you never knew where you would end up. Often we would disturb a large mud eel and pursue him down the drain like a gang of schoolboys, armed with our shovels, yelling and shouting.

We had to work to specifications as laid out in the agreement, so much at so much per yard. As any man who has been on this kind of work knows, every shovelful is saturated with water and as heavy as lead, and often entwined with roots of dead manuka or kikatiha logs and roots which have been buried for untold generations. This timber never rots, as the peat and water seem to preserve it in its natural hardness. In fact, it seems to become more brittle in this pickling process. This invariably has been the case with every extensive area of virgin swamp land I've struck in the North Island. When first viewed in its virgin state it looks as level and smooth as a billiard table, but you don't see the jumbled chaos of the buried forest hidden underneath.

Just imagine standing on the edge of a swampy lake or sluggish-moving river and, as far as the eye can see, looking at nothing but swamp and sky, and miles away the surrounding foothills, and nothing but sluggish brackish water in which you could well disappear when crossing. Believe me, I've seen a fifteen-foot length of manuka shoved in without touching the bottom of this kind of morass. And I was working on this from the bottom of a formed drain, six feet by four feet.

To remove timber stopping the water from flowing, you blasted it out with gelignite. Often you had to cut through a slimy, mud-encrusted log sometimes three or four feet thick, half submerged to your middle in black peat and water.

So progress was often slow and tedious, and a party of three men would toil like slaves to do a chain of this drain in a day, often striking, as well as the logs, enormous stumps with deep penetrating roots reaching far down into the bottom and sides of your precious ditch.

These drains over the Hauraki swamplands had to be kept as straight as a rifle shot. Another thing too: you had to make sure the ditch was graded absolutely correctly so that the fall was right all the way, so that the water at the foothills from which it flowed was drained into the rivers and so on to the sea, drying out the swampland. It was a long and tedious job which took years to accomplish, and I always maintain that it takes three generations of man to bring swamp into full production.

I mentioned how blasting out logs and stumps could block the progress of the contractors. This blasting blows a huge hole in the drain. As soon as the charge exploded, if luck was on our side, it would blow the timber high into the air and on to the surrounding land, out of our

way. But many times I have seen a big stump fly straight into the air after the explosion, then land fair in the middle of a finished section of our work. This then meant hard bullocky work hauling the stump up the drain bank with one chap underneath half bogged, and mates on top heaving and tearing their hearts nearly out. Imagine the stump or log weighing several hundredweight and saturated with water and covered with black slimy mud, being manhandled out. For this strenuous toil we received, if we were lucky, one shilling per yard of progress, or twenty-two shillings per chain of completed drain. Even so, the landowners of Hauṛaki tried to beat the drainers down in their price per yard.

The first farmers to move into the swamplands once the drains were well under way had their problems and troubles too. Hardy souls, these men and women, through grim necessity. Many kept flat-bottomed boats by the verandah, because whenever it rained hard the Waihou and Piako Rivers covered the country for miles around, isolating their homes.

At one settlement on the upper reaches of the Piako River, men from the 1914–18 war were settled by the Government on their savings. The houses were originally built on piles driven into the peat, which had not settled and was saturated with water at every spring tide. The houses gradually sank until the floors rested on the ground. It was quite a common sight to see homes tilting over if one side sank quicker than the other, and in some houses it was not possible to close the doors or windows. How uncomfortable these homes must have been, and so cold and damp in the winter months. Many wives of the returned servicemen tolerated this for a while, then packed up and left their husbands to farm on their own. It was amazing that people stayed on.

But they did. Today these Hauraki Plains, once a floating morass, provide homes and a living for thousands of people, with thousands of acres of rich pasture land, some of the best in the world.

THE LAND OF SACKS

Amelia Batistich

IT MUST HAVE seemed a long way to Dalmatia, up there on Parenga in the godforsaken north. Your house a contrivance of sacks with tin chimney attached, your bed four posts and more sacks, and your days and nights nothing but work, work, work. Never a holiday, or a saint's day; not even a Sunday. Only the gum that must be got out of the ground—the gum that must be scraped and cleaned and sacked; and God help you if you couldn't find enough of it to get you English pounds to send back home.

Work like hell and get out of it, back to Dalmatia and the grapes and olives and the smiles of your own people that say you are not a stranger, that the bread you eat is your own.

And work like hell they did, those foreigners. The New Zealanders called them Austrians, and that was the stamp-mark on their passports, but these subjects of His Imperial Majesty Franz Josef, Emperor of Austria, felt about as much Austrian as the Irish did English.

They were a rough lot. Men without women from a land where a man never did the woman's work, though the woman did both. Here they must cook and wash and clean and mend for themselves. Clumsy fingers grew used to needle and thread—a wonderful array of patches paraded off to work! Those knees, constantly bent to shovel and pick and spear, took heavy toll of the toughest working pants. And a patch made of sacking saved many a skin. No wonder that returning emigrants remembered New Zealand as "the land of sacks".

The land of sacks: its legend lived on in Dalmatia. A little girl used to listen to her grandfather's tales of some terrible place called "Billygoat"—a gumfield on the East Coast. When she grew up she came to New Zealand and laughingly told me how the old man would say to his sons: "When you curse a man, don't wish him to the devil. Wish him to Billygoat!"

"The lonely hell of gumfields." That was how the young Ante Kosovich, poet and gumdigger, described them. For the men who lived on them, the camps were at least places where they shared their exile, and an exile shared is an exile lightened. Many of them were hardly more than schoolboys. My father was sixteen when he came. The idea was to escape conscription. Three years, or four in the case of the Navy conscripts, was a big slice of your life to offer to an unwanted Emperor in return for bed, board, and a pittance of pay. Peasant families could ill afford this drain on their manpower, so boys were sent off in the charge of neighbour or relative to earn money for those at home and help themselves as well.

It was money, or the lack of it, that sent them off to New Zealand. They were part of the great European tide of emigration that flowed out to the new world: an earlier generation had gone off to America. At home they thought it was El Dorado the lucky ones were going to. But the emigrants knew better. "For a dollar a day we built America," one of them wrote.

For the equivalent of a dollar a day, if they were lucky, the Dalmatian gumdiggers helped to swell the earnings from the Kauri gum industry that built more than a few fortunes for the merchant princes in Auckland and earned millions for the young colony.

It was honest money, hard earned. The Dalmatians lived in camps, always staying together, moving off together to another gumfield when the gum gave out. They lived off bully beef; wild pig when the hunting was good. They ate it with camp-oven bread, washed down with billy tea. Their stores they got from the local storekeeper, given on credit till pay-day came round. A faded old exercise book I have seen kept the records of one old gumdigger's simple needs: "Melrose tobacco, one shilling. Camp oven, four shillings. Packet of candles, sevenpence . . ."

Flour, olive oil, yeast, epsom salts (cure-all for everything); these were the staples on the grocery bill. They didn't care much for "Englishman's" butter; in fact, they gave the New Zealander who ate it a name of their own. *Maslar,* meaning buttereater, still used today.

The storekeeper would make out his bill and ride up to the camp on settling-day, for he often acted as agent for the Auckland gumbuyers as well. Then he would present his account, and the weighing-up would follow. Some, it was reckoned, were not to be trusted: how was it that the bill was so much more and the weight of the gum so much less? It's likely that the unscrupulous made a good thing of the foreigner's ignorance of English, but many a storekeeper befriended the Austrian gumdigger, giving him credit where he would not give it to his own.

But trouble was brewing. There were plenty who did not like the foreigners encroaching on their preserves. They saw them coming in droves, settling like locusts on the land. Moreover, it was well known there were only very small pickings for the lazy fellow who didn't want to dig for his gum where the Austrians had been. They left little lying about, and the drifter and the remittance man could eke out a living from the broken-off pieces of gum called "nuts" and take life easy where the Austrians had been digging. Letters began to appear in the newspapers. Feeling ran high, so high that they talked about it in Parliament and a Royal Commission was set up to inquire into the kauri gum industry.

It must have looked bad in 1898 when foreigners without property qualifications were barred from the Crown gumlands.

"Foreigners means Austrians," wrote Ante Kosovich at the time. "Make no mistake about that."

Their friends the storekeepers rallied to their defence. Only one could say that an Austrian had ever left him owing anything. Eight shillings was the debt, freely forgiven because the man had had bad luck.

Well, the New Zealanders might have had some justice in their cause. This was their land, won for them by their fathers at some cost. "We shed no blood for it," says Ante Kosovich, seeing their side of the case. The memories of the Maori Wars were still fresh. But the Maoris themselves liked the Dalmatians. *Tarara* they called them, for the sound of their speech. Many had married Maori girls; there's a fine blend of both races in the North today. And the Dalmatian gumdiggers felt a kinship with the Maoris; they did not feel foreign with them.

Up to 1898 the Dalmatian gumdigger was no more than a bird of passage. He was here to get what money he could, then go back home. But, barred from the Crown Lands, he had to find some other means of getting a living. Gumdigging was all he knew that could be turned to account here. So the first few began to buy up cheap land. When the gum was cleaned out the land itself remained; land that they owned.

That was the turning point.

A peasant people, reared on the stony soil of Dalmatia, the Dalmatian gumdiggers were not daunted by the seemingly impossible prospect of farming the barren gumlands. Married men began to send for their wives, and heaven only knows what tears were shed when these pioneer Dalmatian women saw where and how they had to live.

By the very early 1900s the first generation of Dalmatian New Zealanders was on its way. The local postmasters must have been posed with some problems of spelling when they were asked to register the births of these New Zealanders. One postmaster I have heard of renamed a whole family of girls to his own taste.

Victoria, Alexandra, Tatiana . . . . those babies, reared in butterbox cradles, were a long way from royalty, and so were the homes they were born into. But they grew up into true New Zealanders, proud today to look the long way back to their humble beginnings on the gumlands.

We have all come a long way since then. But the Dalmatian story is written for ever in the history of North Auckland, just as the Dalmatian lineaments are written on the faces of many of its people. Looking back, my father remembered his years on the gumfields with the nostalgia of age for the places of youth. The hardships had all gone, and the gumfields were enshrined in his memory. The old mates were fast going, too, but we knew all their names. Of the New Zealand word "mate" they had made a word of their own, "*met*".*

"*Mete*" they always called one another, and they loved to recall the gumfields they had worked on. Parenga was always the starting place, and the stories ran on endlessly.

"Listen," he would say to me.

And I did, watching the way his eyes shone when he spoke. That big strike, when the probing spear struck northern gold and they were in, gumboots and all, to wrest the last shilling's worth of booty from such luck.

That mysterious find at Houhora, where the land was hollowed for the full length of a giant kauri and the spear hit something hard as rock. What else could it be but gum?

"Don't touch this place," their Maori friends warned them. "This ground is tapu."

Was it the sacred buried canoe? Oh, the thought of the gold that might be hidden there itched at them for a long time, but no one dared to disobey. For the old time Dalmatian had a healthy respect for ghosts. He remembered the stories told at home in the long winter nights when the family huddled close to the stove for warmth and the blood of youth froze. Older men told the same stories around the campfires in New Zealand, with a few Maori ghosts thrown in as well, and many a youth went shivering to his sack bed to listen to the morepork's call falling eerily on the night.

It was all a long time ago. The third generation of Dalmatians is well on its way now. You wouldn't know them in a crowd.

But there's a feeling for the old country yet. Grandchildren of the gumdiggers, dressed in traditional costumes, dancing the kolo at Yugoslav gatherings; they might be, as one New Zealand writer observed to me, New Zealanders in fancy dress. But get that *tamburitza* music going, and something wakes.

For how much longer?

* Author's note: You are my *met*. But if I address you, it becomes *mete*, with the *e* sounded as in *pen*. Most of the immigrants came from the province of Dalmatia, then in the Austrian Empire, but since World War One part of Yugoslavia.

Tell Me a Story

AND THE WIDOW'S EYES WERE GREEN

Arthur Davis

THE FOG CLOSED round them like a white blanket as the two Sams groped their way towards the wharf.

They had been late in with their boat that afternoon because Sam the dog had died at sea. Since he was a sea dog, they had buried him out in the estuary just off the big sandbank where the current along the shore meets the flow of the river and the water swirls continuously.

The truck driver from the retailer had been quite grumpy because he'd had to reverse the length of the wharf to pick up a few bundles of flounder which constituted their catch. He moaned about the fog, too.

After a hurried meal at their hut, the two Sams strolled up to the old Salutation Hotel to give the dog Sam a proper send-off. It turned out to be a night of remembering.

Old Sam bought the first jug, and they remembered the dog Sam.

They recalled how as a pup he had given up snapping at the corks when the nets were running out, after he'd got his teeth caught in the mesh and himself flicked into the sea a few times. That, they thought, was clever. A lesser dog might have gone on doing it for years.

And then they thought of the time he had rushed into the shallow water and dragged ashore a big flounder, single-handed. After that, although they thought long and deep on it, they couldn't remember another single clever thing the dog Sam had done. But still, they thought it was probably only a matter of opportunity. Dogs on boats don't get many opportunities to be clever. Anything they try usually lands them in the sea.

Come to think of it, humans are like dogs in this respect too, reckoned the two fishermen as they drank again and the fog piled up in the night outside the little hotel. When you think of it, some humans get all the breaks while others seem to have the dice loaded against them from the moment they are born. Before they are born, some of them.

Like the widow Malone, for instance. Yeah. Through two whole jugs the two Sams remembered the widow Malone with her red hair and her flashing green eyes. They recalled her hands, forever red and chapped through doing other people's washing; her numerous brood;

and her fierce spirit of independence. She could smell charity a mile off, and insisted on doing their washing and mending in return for the fish they brought her every day. On Sundays they put on their best bibs and tuckers and went to the widow Malone's for their tea.

They remembered the widow Malone's family growing up and leaving her one by one. When the last one went she seemed to give up. She wasn't the same woman any more. And then they recalled the day they had taken her to the hospital. They had carried her in through the front door, and after a few brief weeks she had gone out the back. And so she never came to claim that pup—the pup she had left in their care and had named after them.

"No," the two Sams agreed, "they don't come any better than the widow Malone—and they don't come often, either."

They briefly recalled her drunken husband who had fallen down a mine shaft and broken his neck. The only decent thing he had ever done; and even then he didn't do it soon enough. You could say that again. . . .

The two Sams grinned as they drank and leaned forward and looked into their glasses; the two Sams grinned as they thought of their first venture together, when as boys they had run their nets over the stern of a dinghy far up the river; and the day they went to Auckland to take over their new boat, a gleaming thing of white paint and polished brasswork.

With the classy new boat they had fished all over the harbour for snapper and flounder; and when the season was right they headed out after hapuka in the deep waters of the gulf.

They didn't go out into the gulf any more.

The old engine wasn't up to it, and a lot of expensive-looking fungus had begun to grow inside the boat along the waterline. Besides, the big boats came down from Auckland now, with their echo-sounders and refrigerators, and took the cream of the catch without even trying. "Push-button fishermen," the two Sams called them.

And so they were earning a living back up the river again. The wheel had turned full circle, and they were battling with the fierce currents where no one else would go for fear of losing his nets on the outgoing tide, or having them filled with the jellyfish that came in with the warm weather, so that the heavy nets would sink without trace.

But they weren't going up the river tonight. No fear. They knew of a little bay away down harbour, ringed in with mangroves and accessible only on the spring tide. They had taken some big hauls there in the past, and at daybreak the following morning the tide would be just right.

When they finally left the Salutation Hotel the fog was thicker than ever. Despite the grog and the fog, they managed to reach the wharf all right. But old Sam had to caution young Sam several times about the danger of falling off, as they made their way along the wharf. They got on board without incident, and sure enough the motor started at the first crank of old Sam's practised hand. After a few minor adjustments it settled down to a hesitant wheezy rhythm, which was about all they could expect of it these days. Young Sam slipped the bowline off the capstan and pulled it on board. He poked his head into the wheelhouse.

"All s-set," he said; then somewhat unsteadily went aft, flopped on the nets and instantly fell sound asleep.

Old Sam pushed the stick forward. The motor faltered slightly as it took the load, but eventually settled back into its uneasy beat. The fog was so thick that even the leading lights were out of sight.

Old Sam wasn't worried about the fog. He knew the harbour like the back of his hand. With his compass and his battered old alarm clock he could find his way anywhere. He knew the set of every current, the depth of every channel, and the bark of every dog along its rocky shoreline. But he had lost a lot of faith in dog-barking navigation since he ran his boat aground simply because the farmer on the point had sold his dog to the bloke in the bay.

He glanced at the clock. Thirteen minutes and they would be off the big sandbank. He would have to cut round it. Seemed to be growing each year. Once you could cut over the top at any tide.

He missed the dog Sam. It was unusual to see the box on the other side of the wheelhouse empty. He had always discussed his navigation problems with the dog, but apart from showing pleasure at being consulted, the animal had ventured no opinions. He wasn't much of a navigator; whatever old Sam decided had been all right with him.

Old Sam turned the wheel a few spokes. He imagined he could hear the waves washing over the sandbank. The dog lay somewhere below them . . . and the widow's eyes were green. Aye. . . .

He now had a straight run of thirty-five minutes down the coast to the point jutting out to sea, guarded by big rocks richly clustered with oysters.

It was a long thirty-five minutes. He wished now that they hadn't stayed so long at the Salutation Hotel. Far better to have shown a bit of intelligence and gone home and got some sleep before setting out. Especially for young Sam, who hadn't weathered the years very well. No one would imagine, thought old Sam, that he was a whole year his junior.

He thought he had reached the point at last and started on the trickiest part of the journey, involving numerous course changes and careful time checks. He was quite worn out when he finally stopped the motor and went forward and threw the anchor over.

He walked astern and sat on the icebox. He couldn't see young Sam, but he could hear him snoring. He had always snored like that. Adenoids or something. Old Sam had advised him to have them out. Young Sam said he would, as soon as they interfered with his swallow.

The mist was beginning to clear. It was starting to swirl as the wind caught it. Patches of open water became visible with the breaking day. Old Sam peered ahead, but he couldn't see the mangroves. He must be getting old.

Then, quite casually, he glanced astern. His eyes widened in disbelief. There was no doubt about it—young Sam was losing his grip.

He had forgotten to cast off the stern line. They were still tied up to the wharf.

WHEN ROSALIE FELL FOR THE HORSEBREAKER

Arthur Davis

THE THREE DROVERS sat in the bar-room of the little country pub. This was the moment they had been looking forward to for the past week: the sheep were on the trucks and the rest of the day was their own.

The oldest drover toyed idly with his half-empty glass.

"The old pub's dyin' fast," he said sadly. "I mind the time when four barmen was goin' flat out at two o'clock in the mornin'. Now look at it. Three drovers and pub's full. The beer ain't as good as it was, neither, come to think of it."

The other two were reading a newspaper they had divided between them.

"I've got all the blasted advertisements," complained the youngest. "Hey, wait a minute! Did you see this? Old Mick Dooley's place is on the market. Seventeen thousand quid he wants for it, L and B. What's L and B?"

"Land and buildings," volunteered the oldest drover.

"Where would a working bloke get that sorta money?"

"You could pinch it. Not a very good way, but it's a way. One thing, you could never save it. P'raps you could borrow it. If you did, you'd find that these moneylenders have a great capacity for looking after themselves. They've been at it for years and getting better all the time. Anyway, no working bloke will buy it, so why bother yer head? Some bloke with tons of money will snap it up. Farming's big business today."

"Someone like that little fat coot in his big Mercedes that went through the sheep yesterday, blowing his horn all the way," suggested the young one.

Funny you should mention him, the old drover went on; knew him years ago. He was only a horsebreaker then.

He was working fer old MacDougal. He was one of the Clan MacDougal, whoever they were. But I do know that they didn't chuck money around in flamin' great handfuls. Still, times was hard in those days. You had to take what you could get. Old Mac hired the bloke you saw in the car to break in a dozen or so horses while we were there.

He came in an old truck with all sorts of gear on the back, including a portable forge and a big pin vice screwed on the tailboard.

Said his name was Alan. Cheerful sorta cuss. Played the mouth organ and the guitar and sang in a deep voice. Didn't look much like a horsebreaker, though. No jeans or highheeled boots. No forty-gallon hat, no spurs. Looked more like a scrubcutter.

We ran the unbroken horses in next morning and each of us drew straws for first pick. That was the rule; once you had your pick, that was the horse you were saddled with, no matter what he turned out like.

The head shepherd didn't draw a straw. He said he'd take what was left. He wasn't a very optimistic sort of bloke. Used to sit in the smoke while he was boiling the billy.

Alan didn't bother to come over. Just told us to pick out four, which was the number he liked to handle at the one time, and let the rest go.

Old Mac told him that when he was horsebreaking he always handled six.

"Just a matter of preference," said Alan. "I like four."

"Well, ye'd better come over and take a keek at them."

"No need," replied Alan. "I'll just check over me gear. Horses is just like people. Different sizes, different colours, different natures; but when it's all boiled down, just horses."

We didn't see Alan make a start on them. Old Mac didn't believe in paying a horsebreaker and paying a lot of onlookers as well. In fact, he didn't like paying a horsebreaker at all. In his young days he said every shepherd worth his salt broke in his own horse. And in his spare time, too. He seemed to overlook the fact that to follow in his footsteps we'd have to have some spare time.

He did go over once himself to give a bit of advice, but must've got the idea he wasn't wanted, because he didn't go again.

Alan had a lot of unusual ideas. He was a neat man with a rope and could tie a horse up in half a dozen different ways.

"It's no use me wrestling with them," he said. "They're a lot bigger than me. And anyway, it's not me they're fighting. It's the rope. I'm the bloke that lets them go."

He never raised his voice. Never did his bun. He made them responsible to him for every bite of food and every drop of water. You know, after a few days those horses used to nicker when they saw him coming. The way he did it, horsebreaking was such an easy job. Seemed a shame to take the money.

We was all waiting for him to ride them, but he didn't seem to be in any hurry.

"Can't risk them buckin'," he explained. "You see, I've never ridden a bucker."

Anyway, they didn't buck. They didn't even kick when he shod them. Real disappointing it was. When we took them over they never gave a bit of trouble. Quiet as sheep. Even the boss's daughter Rosalie could ride them.

We always wondered how old Mac and his wife managed to get such a goodlooking daughter as Rosalie. We could easily understand where she got her dirty disposition from. The old lady had a bite like a "blue-heeler".

Almost everybody had given Rosalie the glad eye at some time or another. She allowed them to escort her as long as it suited her, and then she knocked them down and ploughed them under when something better came along.

The worst of it was that she was pretty enough to get away with it. It wasn't long before she had Alan lined up in her sights.

She got old Mac to take him off the horsebreaking to give her a hand getting her pony ready for the sports. He spent days oiling her gear until the leather was soft as cloth, and then he sealed the oil in white of egg so's the oil wouldn't get on her britches. He polished buckles, bits and irons until they shone like silver. He brushed and scrubbed the pony until it was almost as good, and even polished the clinches of the shoe nails.

He went further than that. He taught Rosalie how to hold her head and hands and how to point her feet. He knew a lot about show riding. He didn't seem to mind his job.

"Anything for a crust," he said.

Rosalie scooped the pool at the sports and her photo was in all the papers. Alan was the snowy-headed boy round the station by now, and when he tied a fly that allowed Mac to catch the big brown trout that he'd been after for years, and got rid of the bugs on the old lady's roses, he could do nothing wrong. He was having dinner up at the house now, and we didn't see much of him.

It was then that we decided to slip the Man-eater into the second batch of horses he was ready to start on.

The Man-eater was a big upstanding horse with a fine head and the body of a thoroughbred. A lot of horsebreakers had had a go at him, and he'd beaten the lot. They broke him in all right, but somewhere along the line he'd learned to bite, and it wasn't no playful nip either. He'd bare his teeth and let out one god-awful scream before tearing a hunk out of you. You never knew when he was going to do it either, which didn't make for dreaming when you was leading him along. Finally he got so bad that he'd chase you clean out of the paddock when you went to catch him.

Old Mac had been going to show us how to cure him, and got himself bitten on the elbow so that he was off fishing for a fortnight. After that he used to say "A good horse ruined" every time he saw him.

We weren't there when Alan started on them. We always seemed to be missing the good bits.

When we got home in the evening, all Alan said was: "That big white-faced horse has been handled before." He then cranked up his old truck and drove away.

He was back in about an hour and set off over to the yard carrying a little parcel. We followed at a discreet distance. When we ranged ourselves round the yard Alan was at the gate of the pen containing the Man-eater. He was busy with a cigarette and the parcel. The Man-eater was facing him with his teeth at the ready. After a bit Alan opened the gate and the Man-eater charged him with the whites of his eyes showing and his mouth wide open. Alan handed him the parcel containing a big cracker wrapped up in cloth, then slammed the gate in his face. Next morning the Man-eater wouldn't bite anything; not even grass.

After he'd finished that four, Alan was taken off the job again to get Rosalie's horses ready for the show. He rigged up a set of hurdles and they set about schooling the jumpers.

Alan knew a lot about jumping horses. He knew when to "check" them and when to "lift" them. Under his tuition Rosalie was coming on fine.

On the day of the show they were away bright and early, and Rosalie scooped the pool again. She won every event she entered, and of course the horse she won the most on was the Man-eater.

At the ball that night Rosalie decided she'd had enough of Alan. The time had come to whistle him up and put him under. But Alan didn't come when she whistled. Instead he went on dancing with the Benson girl, singing in his deep voice and showing her all the latest steps.

It took Rosalie a fortnight to get him back again.

Which goes to show that humans is just like horses. Different sizes, different colours and different temperaments, but after all just humans. You just gotta know the answers.

Alan doesn't break in any horses now. Neither does he sing or play the mouth-organ. He's too busy with his stud stock and with piling money into the bank, where they'll stow it away and he'll never see it again. Since he married Rosalie and they inherited the station, I guess he can afford to drive through mobs of sheep in a big Mercedes, blowing the horn all the way.

So that's another way you can get money. You can marry it.

But I think if you was to ask Alan, with his wife the dead spit of her mother, he'd tell you that you can borrow it cheaper.

GIGGLING PIN VERSUS THE BOSS

Arthur Davis

IT WAS BOXING DAY up at old Ardgowan and already most of the shearers had arrived. They'd come in a variety of vehicles, mostly in advanced stages of decay, that had crept wearily down the hill from the main road, spluttered hesitantly through the ford and finally drawn up exhausted before the shearers' quarters.

There was Scotty Fraser who had begun shearing there with the blades and had never missed a season since, and Baldy Thomas who had a bit of a run up in the back country somewhere where he ran a few sheep and several million rabbits. There was Tumbling Tommy from down Southland way who came up every year, to get away from his missus, they said, and three young fellows who had just cut out a run up in Nelson where they said the sheep were tough.

And then of course there was Giggling Pin with his tin mug and his mouth-organ. On top of these there were the roustabouts and the shed hands and the inevitable penner-up with a downcast dog on a string.

Late in the afternoon a taxi drew up in the yard and Happy Jack descended from it. He settled like a cloud among them. Happy had been everywhere and he knew everything. By nightfall he had had a row with the cook, spoken to a shepherd about some barking dogs, discovered that the shearers' quarters were nine feet closer to the woolshed than the regulations allowed, and that the cubic airspace per man would need further investigation.

He had also "chipped" Giggling Pin for playing his mouth-organ. Poor harmless old Giggling Pin. All he wanted to do was play his mouth-organ and tell of the time he had rung the shed, up at Haka. Not that anybody believed he had rung the shed at Haka, or anywhere else for that matter. All he seemed interested in now was in shearing about a hundred sheep a day and making a perfect job of every one of them.

Happy Jack kept wandering about making his complaints to anybody who would listen but as yet he hadn't met the boss. None of them had. It was only recently that the new owner had taken over from old

Jock McNab who, tired of the high country and his burden of years, had retired to Timaru where he aimed to play bowls.

Of the new boss Happy Jack spoke darkly: "He'll have to be eddercated," he said.

The cowman-gardener, when approached, was inclined to disagree. The boss, he said, was already so well educated that you couldn't understand a word he said.

Actually they didn't meet the new boss till the following morning, when he appeared on the board just before the gong.

"Make a good job of them, boys," he said in his affected voice. "No wool on the sheep and no skin on the wool."

Happy Jack winked at old Baldy: "And what would you call a good job?" he asked innocently.

"Pink 'em, old chap. Pink 'em."

"Well," said Happy. "Suppose you shear one, just to show us how it's done."

The boss looked down at his jodhpurs and his brown calf boots.

"Well, old chap, I'm not exactly dressed for the part."

Happy Jack was openly sneering now. "That's the trouble with you cockies. You're all the same. You want a first class job and yet you can't show how it's done."

The boss never said another word. He just reached into the catching pen and dragged a sheep on to Happy's stand.

Three swift blows and the belly wool was gone. He was round the crutch, over the hind leg and up the neck before they knew what had happened. Round the head with not a movement wasted. Down on to the first shoulder in a flash, and then on to the long blow. The boss really poured on the speed on the long blow. Then he pulled its head up between his knees and wiped off the last side with long rhythmic blows.

"Just do them like that," he said as he pushed the sheep out the porthole.

The shearers looked out their windows at the newly shorn sheep. Not a scratch marked its skin and not a tuft or ridge of wool masked its pink nakedness.

The boss put the handpiece down.

"That comb is far too sharp for fine-wools," he told Happy. "You'll cut them to pieces with a thing like that."

He walked down the board and out into the yard.

"A minute and a half. Not bad for a bloke with a voice like that," said Giggling Pin. "I thought he mightn't be all he sounded. I had a look at his claws when he walked in."

"Well he knew how to pick a good one," said Happy Jack sourly.

Happy was a broken man. He knew he could never shear sheep like that and make money. He did, however, make one final effort to save his face. He complained to the boss that the first-aid box wasn't complete.

"It was complete when we started," said the boss. "You must have been using it to repair the frightful gashes you have been putting in your sheep."

After that nobody was surprised when Happy Jack was suddenly taken ill and departed in the same taxi that had brought him.

"He'd been edder-cated," summed up Giggling Pin.

His stand stood empty through the shearing. The mobs of sheep came and went until finally there was only the last block to be brought in. The shearers were worried about this last mob. They inquired its numbers from the head shepherd.

"Fourteen hundred," he said. "And you can take it that the tally will be pretty right. That's the number that the boss bought off old Jock McNab, so he'll be pretty right."

The mathematicians went to work and very soon found out that on present form they had no hope of cutting them out the following day, and the day after that was the country race meeting down at the settlement.

They looked anxiously out on the ranges for any sign of rain but the day was as fine and as hot as the ones preceding it. Finally they saw the last mob coming round the spur, and the boss's stocks slumped quite a bit.

"Old Jock," they said, "would never have sent his musterers out when it was going to mess up the races. A thoughtful man was old Jock."

By nightfall they were resigned to the fact that they were going to miss the races. At first they thought that if they went flat out they might be in time for the last few races, but finally decided it wasn't worth the effort.

Next morning when they arrived at the shed they were surprised to find a worn old handpiece set up on the empty stand and the boss down in the wool-room making himself a pair of sacking moccasins.

Everybody cheered up immediately.

"If he opens out like he did on that first sheep," said old Baldy, "nobody'll be able to keep him in sight. You'll have to stay with him, Gigglin' Pin. Shear like you did when you rung the shed up at Haka."

Giggling Pin said nothing. He was old now. Many years had come and gone since he shore up at Haka; but come to think of it the boss was no chicken either.

"A man can but do his best," he said finally.

Nevertheless everybody was surprised when he and the boss let their first sheep go together. The next was the same and word went round the woolshed: "Old Giggling Pin's having a go at him."

They raced neck and neck through the first run with neither yielding an inch. Their pen doors banged together with monotonous regularity until finally the boss began creeping ahead. By the end of the run Giggling Pin had lost a sheep.

In the second run, however, all the short cuts and the fancy blows old Giggling Pin had forgotten came back to him, and he managed to get it back. So at lunch they were level once more.

That afternoon the nor'wester sent its hot breath down from the hills and filled the shed with fine dust.

The shearers toiled on. The boss and Giggling Pin were still dead level and all the other shearers were turning out tallies that they would have reckoned impossible the night before. The rousties flew up and down the board while the table-hands tore at the necks and skirtings the moment the fleece was thrown. The penner-up rushed from pen to pen banging his gates up and down with a tremendous clatter. The old dog became so excited at his shooing and shooshing that he gave several hysterical barks and then bit the penner-up. The wool-classer lost a lot of his polish under pressure and swore unpleasantly at the woolpressers for branding a bale wrong.

At the end of the third run it became obvious that the sheep weren't going to last all day. As their numbers grew gradually fewer, the musterers came into the shed. An early cut-out meant that they could get the sheep back to their paddock that evening, and go to the races as well.

Giggling Pin and the boss raced neck and neck until only one sheep was left. The other shearers slowed down. They didn't want it. They left it to decide the contest.

Blow for blow these two raced over their sheep until they got on the long blow, and here Giggling Pin struck trouble. His sheep started to kick, and nothing he could do would stop it. The boss raced away. Old Giggling Pin was beaten. A pity, after the show he had put up, but there it was.

The boss was casually finishing off his sheep when the learner who nobody had thought about walked down the board, picked up the winning animal took it away and shore it.

And so it ended in a draw.

That night the boss brought over a barrel of beer in the Land Rover and stayed at the quarters until it was empty. Old Giggling Pin played his mouth-organ and when the beer was finished went to his bag and brought out a bottle of whisky reverently wrapped in a singlet.

Old Giggling Pin doesn't shear any more now, nor does he tell of the time he rung the shed at Haka. Instead he tells of the day "me and the boss shore pen mates up at old Ardgowan". The boss doesn't shear any more either. In fact since he made Giggling Pin boss of the board he hardly ever comes near the shed.

When he does he and old Giggling Pin sit on the bales and he tells tales of faraway places like Maru Narribri and Yass, where the woolsheds are of palings and there are three curls in the Merinos' horns.

Old Giggling Pin listens with a distant look in his eyes, and in moments of fancy, imagines that perhaps he too might have shorn there as well.

"Because It Is For The Community"

A HARD DAY'S NIGHT WITH THE POWER BOARD FAULTMEN

Trevor Meyer

THE RAIN'S LASHING against the windows, driven by a wind that's shrieking like a mad woman, and through the closed blinds stabs of lightning let me know the thunder's following. I'm awake listening to it, and as I think of the call that's bound to come, I pull the blankets tighter around me.

Sure enough the 'phone rings, and even in the house it feels damp and cold as I climb out to answer it.

"Hullo," I say into the mouthpiece, a bit surly like from being dragged out of bed.

"It's Colin here, Bill. There's a job for you out Okaihau East. Four reports from the same locality. A transformer, most likely. Come up here and pick John up as you go, eh?" Colin's the senior faultman in the office.

"OK," I say, and plonk the receiver down.

The power is still on, so I plug the jug in to boil while I dress. It's a long time to daylight, and heaven knows how many calls we'll have before then. Nothing like a good cuppa to set you up before you leave. Over my dry clothes I drag the cold leggings and oilskin coat I'd been wearing the day before, shove my feet into the short gumboots, and knot the sou'wester under my chin. Then out the door, and battle against the wind to reach the truck.

The wipers work overtime as I head for the office.

"Nice night," John greets me. "Trouble everywhere. *And we won't be home till morning, and then we won't be home,*" he sings cheerfully, as we set off.

The R/T crackles as we drive along, and I call back to the depot to check reception. Blackness all round us except for where the lights cut a track. We get spasmodic glimpses of the countryside as flashes of lightning seem to tear open the clouds and let another deluge through. Can't hear any thunder for the noise of the wind roaring past the truck. Then the tarseal ends and we jolt over the potholes on the metal road.

We reach the transformer where the faulty line branches off the main lines north. The spotlight from the truck shows everything OK there, so we're off down the road to the next tranny, which feeds two farmhouses now completely in the dark. We pull up with the rain bashing in bucketfuls against the truck, and the wind shrieking, with extra gusts rocking us for good measure. There's no sense in waiting for a break, it could go on all night.

The wires gleam whitely as the spotlight picks them out near the top of the pole. Something crunches under my boot on the roadway. It's a glass tube with a brass end: that's the fuse and it's well and truly blown. In theory, according to the regulations, a man isn't compelled to climb a pole while there's lightning around. But hell; there'll be a list of faults a mile long by now. Might as well get cracking and do the job.

We offload the ladder, prop it up against the pole and pull the extension cord to make it the right height. John climbs up it carrying fuses and fuse stick. The pole sways in the wind, shaking the ladder and making it hard to keep balance. He feels much safer once he's secured the top of the ladder to the pole, and fastened his safety belt too.

Not a place for taking chances, up there.

The ground's a long way down, and up above you there are live wires. John puts on his gloves and goes to it putting the new fuse in. Looks simple from the ground. Just push the fuse in the fuse stick into the clips on the pole. But up there a man could do with an extra pair of hands, what with trying to keep his balance, and match the fuse up with the clips on the swaying pole. Not easy, but he does it, and out of the darkness to our right comes a yellow square of light in the farmhouse window. That's worth something anyway: we don't need to muck around going into the cowshed I knew was handy, or knocking anyone up to check if the power's working.

John undoes his belt, then the ladder, and climbs down to load the gear back on the truck. The warmth in the cab makes him shiver; it's been pretty cold up there.

Half a mile along the road I go through the drill again, and as John remarks: "There's another bunch of satisfied customers."

We wait in the cab for a break in the stream of messages on the R/T, then I report our faults cleared.

"OK," Colin acknowledges, and sends us on to Waipapa, about half a dozen miles further on down a road more like a stream-bed by now than a roadway.

A transformer has received a direct hit by lightning and burnt out. The distorted tank gapes open at one seam, and dark oil drips out steadily to be blown away on the wind. Nothing we can do with it

7·5KVA

except isolate the fault by removing the fuses further back. A new transformer and daylight are necessary to fix that job.

Another session on the radio, more jobs to do. Some more fuses, and a tree across the wires. These trees—won't some jokers ever learn they don't mix with power lines?

Toward daylight we pull in by a cowshed where a cocky's waiting with a yard full of cows for the power to come on.

"You blokes enjoy your sleep?" he asks sarcastically. "I've only been waiting over an hour for some power to get the milking done. A man shouldn't have to wait like this. After all I do pay for it, and it ought to be here when I want it."

We've been on the go five hours now, and not in the mood for remarks like that. But what's the use of arguing with the old coot? We mutter to ourselves, ignore him, stick the fuse in, and take off, still resentful.

But a few jobs later we feel OK again.

"You chaps do a great job," the cocky says. "Damned if I'd hare round the countryside in this kind of weather. I wouldn't go near those lines in a thunderstorm for a million quid. Could you go a cup of tea before you move on?"

Could we? It's been a long time since the last one, and a hot cuppa with biscuits warms us up and makes us more perky.

Next call over the R/T we tell them we're nearly out of fuses, so they tell us to come in, and if we're lucky we'll get a bit of a break.

But just as we strike the tarseal, a voice calls us on the blower.

The storm has moved back into the area where we did our first call, and the power is off again. Could we do it while we're out that way?

Curse the stinkin' storm! Once in a day is enough to fix any fault.

We find a gateway, turn the truck and head back. It's mid-morning, but still murky under the dreary rain. As we pull up at the pole there's a blinding flash and simultaneously an explosion of thunder that seems to compress the truck cab. That was near! Torrential rain pelts down like machine-gun strafing. Where the devil does it all come from? It hasn't let up all night.

My legs feel a bit wobbly as I climb the ladder. It's not a new exercise, but it's getting a bit overdone today, and I'm cold into the bargain. The lights flick on and off as I start down the pole and the put-putting of a vacuum pump comes brokenly on the wind. Another one done.

The truck splashes through a stream where it has overflowed across the road since we came down, but we're on our way home now.

"I could really go a good hot feed and I don't care how soon I wrap myself around one," says John.

"Me too," I tell him, and my stomach quivers at the thought of it.

We arrive back at the depot and are given an hour off before reporting back. Everyone they can rake up is out on the faults; it'll be a miracle if they're all fixed today. This storm has cost the Power Board a few quid.

It's only an hour, but there's time for that feed, and a smoke that hasn't been dampened by rain. Then it's back to the job.

The rain has turned off like a tap, and the sun shines weakly. But the breeze has an edge like a knife, and cuts right into our bodies tired from lack of sleep and battling with the storm. We get chilled right through; worse than when it was raining. But the last fault is fixed at last, and we're thankful to head into the depot.

"What made you take this job on?" John asks me wearily.

"Dunno," I answer, thinking about it. What makes a bloke milk cows or push a pen in an office all day? "Damned if I could stand being cooped up in an office all the time; like a prison."

"Nor me," John agrees. "I like an outdoor life. But I've sure had an overdose of it today."

COUNTRY DOCTOR

Anonymous

Over the hills and far away drives the country doctor, blessing the milk tankers which made cattlestops necessary: no more gates to be opened or shut in the rain. Little houses, big houses, the high and the humble; up the drive or up the muddy path (never run over a dog or a cat or a chook, and watch out for the pet lamb)—here's the country doctor on the job:

Doctor: These people aren't in such a hurry as they are in the cities and I think they're healthier. They have more accidents but I don't think they get the same type of illnesses as in the cities—or not so much anyway. The great thing about a practice in the country is that you know more about the people; you know their families, and I think you can understand their illnesses much more because you know them so well: their children, their parents. I've known some of them for twenty years and there's not much about them I don't know.

Do you think a special temperament is required for this?

Oh yes, you have to like your patients. I s'pose I like most of mine. But if I didn't know the people so well, I wouldn't be able to treat them so well.

What's the difference between a country practice and a city practice?

I think I do more visiting. I think too, that I go into the people's homes more, and I can understand how they live and why they get various illnesses more than I would in the city, where I wouldn't ever go into their own houses; they would visit me more there.

Do you envy the town doctor at all?

No, not really.

A tiny house of an old settler, the roof red with rust, a large white camellia alongside; a State house, a lonely pensioner's cottage; a lonely pensioner's spick and span two-room flat; a split-level house with a thirty-year mortgage and a garden looking like an architect's idea of a garden; a skew-whiff house with the tank-stand nearly falling down —up to the door goes the country doctor with his little brown bag:

Oh yes you do everything—or try to do everything. The things you can't do you refer to the specialist, but you see most things in medicine.

What about the misery of phone calls in the night?

I don't like them, but then I don't really mind them either. I'd rather do obstetrics at night because then they won't interfere with the work during the day. But most people don't call you in the middle of the night unless they're sick, and then you know you have to go. If you don't go, you worry about it and you don't get any sleep anyway.

You have a fair bit of paper work?

Yes, a lot which I dislike; but it has to be done. It usually piles up and then there's a few nights' work, to catch up at the end of the month.

And within all these scattered homes, sick children, a sick wife worrying about the milking; a brave old Digger with his metal leg, his souvenir from France ("The trouble was it was a French bullet," says he, sitting by the kitchen stove as the doctor feels his pulse); a man in a coma; a man devoted to Latin and Greek; an elderly woman all alone and remote, in bed with a red shawl round her fine old face, reading a book about bullfighting—and also a book on her tribe, the Arawa tribe. The doctor fills his pipe and asks: Is she sure she's still got plenty of good fresh eggs for breakfast?

Do you run into much loneliness?

Yes, old people get very lonely and they're the ones who get sick because they get lonely. Even if they have relations around them, they don't seem to see enough of the people they want to see; when they've got too much empty time they get sick. If people tell you what they're worrying about it helps a lot. Sometimes they won't tell. But you must advise them and I think when you do and they take your advice it relieves them, and they seem to get over their worries.

He's the only doctor in his country area, fifteen miles by thirty miles, with a population of 5,000. He's been here twenty years. He just doesn't get time to get sick himself. Each year he drives 18,000 miles. Each year he brings about eighty babies, eighty brand-new young New Zealanders, into the world:

The best is probably obstetrical, and of course seeing patients you like. I think obstetrics are most satisfying; very occasionally women are sick when they have a baby, but mainly you're dealing with healthy people, and once the baby arrives everybody's pleased.

What about the sudden call to a serious accident?

Well, you usually think the worst, but most of them aren't as bad

as you imagine—some bad ones and some—well, you get the ambulance—some people are dead, some can be helped; but it doesn't happen so very often.

How on earth do feelings and skill fit in, in an accident? Would too much sympathy affect you technically a bit?

Well, you don't feel any sympathy at the time, because you're busy; later, when you think it over you feel sorry for the people, but not at the time.

Up about half past six, read the paper, breakfast over by eight o'clock and then off to see patients until about ten o'clock. Usually see six to eight patients, travelling twenty or thirty miles before ten. Morning tea, then surgery 10.30 to 12.30 or after. Half hour for lunch. More calls till two o'clock. Surgery again until a quarter to four; then out and away to more calls till half-past five. Dinner, then surgery 6.30 to 8 o'clock on Mondays, Wednesdays and Fridays. After that, an occasional call at night. Work on Saturday mornings for about four hours, a few calls on Sunday as well—altogether about a sixty-hour week.

Oh yes, I get very tired sometimes but one good night's sleep usually puts me right.

Do you worry at all in bed? Can a case, a person's predicament, run round in your head?

Oh yes, especially with children. Children are more worrying because they can be very sick, or appear to be, at one time and be quite well the next day perhaps. But I lie awake and worry about children quite often.

In some country districts the people are convinced that the local vet has fewer animals to look after than the local doctor has human patients. Why is there this shortage of country doctors?

I think because it's a lonelier life. These days too the tendency is for group practice, and group practice is only possible in towns. Where a man can share his weekends, his weeks off, nights off, it's very much easier practice. I'm on all the time. Most young practitioners now don't like that.

"The main thing," says the country doctor, "is to get people well, and if they're well, they're happy. Young people are probably the most rewarding to treat, because when they're well, they know it; they don't imagine things."

As you travel with him, you sense something more: a feeling for humanity, a love for his patients; it must be there; and so, with all his experience, what does he think of life and death?

Well, that's difficult. I don't think people mind dying as much as they say. People always like to live, but when it comes to the point they know they're going to die; you don't need to tell them. They seem to know, themselves.

And sometimes a peace or a tranquillity can come?

Oh yes, most people die very quietly and very easily, and they don't seem to be frightened; they just die.

Do you feel it's a challenge in a way—are you a sort of adversary of death?

You have to keep people alive, and you keep them alive until you know they are dead; it's a thing you have to do.

You must have seen great strength or will-to-live in some people?

Oh yes, some people live on for weeks and weeks when there's no real hope that they ever will recover, but they're determined to keep on living.

And, from all this, what's the biggest reward?

Well, that's very difficult to say. But I suppose if a person is grateful—and most people are grateful but they don't show it—but when a person's grateful for what you've done, that is rewarding.

THE COUNCIL MEETING

Trevor Meyer

WHEN YOU DRIVE round the metalled roads of our countryside and bounce through the potholes and across the watercourses where the water has run because the ditches have been blocked, and in summer the dust comes up in choking clouds, and in winter the slush makes a dirty coating along the sides of your car and piles mud in thick layers under the mudguards; don't you begin to mutter that it's time they did something about it? And by "they" you mean the local body—the county council. What do you pay your rates for? What do you elect them for?

Every bump you strike triggers off a bit more indignation, and you swear the first thing you're going to do when you get home is ring your riding member and really tell him what you think of the state of the road.

Anyway, that's how I felt as I drove our new car home for the first time. So I rang up my riding member and dragged him away from the nice fire he'd just settled himself in front of after getting all his outside jobs done. I told him it was a dirty night (which he knew already), and after a bit of chat got round to the reason for my call. To my surprise he cheerfully agreed that the road was shocking and that something should be done about it. In fact, he had already brought it up at a previous council meeting, and it was down for consideration on next year's estimates.

"Tell you what. It'll be discussed at the next county meeting. Why don't you come along? The meetings are all open to the public."

I half-heartedly thought I might go along; and left him to go back to his evening peace till some other ratepayer hauled him to the 'phone. Night time is a pretty good time to get hold of a county councillor—he's usually busy on his farm all day, and if you don't ring him during his meal hours or at night, you miss out. It's one of the joys of being a councillor.

To make sure I didn't forget, he rang me the night before the meeting, and so the next morning off I went to the county office. Very few people exercise their democratic right of attending council meetings, and there was only one other member of the public there besides myself.

There was a T-shaped table in the long room, and at the top of this sat the county chairman with the county engineer on one side and the county clerk on the other. Down each side sat the councillors. Over near the wall were a few tables and chairs for the press. Back from the table, where they could have a good view of the proceedings, were a few chairs for the general public—the whole two of us.

On a table inside the door lay a pile of cyclostyled booklets, and seeing the other chap had one I took one too. While waiting for the meeting to begin I thumbed through it. First was the Agenda; that took up half a dozen pages. The next page and a half were taken up with the Supplementary Agenda. This was only the start. There were about fifty pages altogether, containing the County Planning Committee report, the Works and Plant report, Staffing Committee report, County Engineer's report, County Overseer's report, Maori Lands report, Contract Tenders for consideration, and local fire brigade business. There were several pages of accounts for payment, too. Altogether quite an eye-opener.

And as I turned the pages I found the chap who represented our riding was on several of the committees and he'd been on inspection tours since the last meeting. I realised this was no one-meeting-a-month job, as a lot of people thought. He'd been on the go quite a bit. And don't forget those telephone calls at night when most men are settled comfortably after their day's work.

Well, the councillors arrived and sat themselves round the table in their appointed places. There was one councillor or riding member for each riding, or subdivision, of the county. The meeting came to order, the minutes of the last meeting, as circulated, were confirmed, and the work began.

"We'll take the Planning Committee report," said the chairman. "It's on page seventeen." Aha. The wheels were beginning to turn . . .

With the rest of them I turned to page seventeen. The items were all set out with the committee's recommendations beneath them. First, a chap who'd sold his farm but wanted to keep half an acre for his retirement; the committee did not recommend the creation of isolated residential sections in rural areas. "Quite right," the councillors said, and, on the voting, the application was declined. Next, a subdivision of property on a town boundary; "OK," said the committee, and council agreed again.

A chap wanted to build a block of flats in town. The committee suggested that further information be given about the project. "Deferred until next meeting" was the verdict.

So it went on, through every application. I began to see the value of the committee. Supposing all these things came before council without some preliminary investigation? How would the bloke from the western riding know what the situation was in the eastern part of the county? And vice versa. The committees save the council from sitting continuously, as they'd have to do if they had everything to decide without the major points being sifted out for them.

"Are there any deputations?" the chairman asked.

Yes, there was a three-man deputation from an isolated corner of the county, asking that the track to their farms be formed and metalled to give them an all-weather road. The spokesman had a few notes on the distance, the number of families, the amount of rates they pay, and the difficulty of getting their children to school.

"Some of us have been paying rates for thirty years. We could have done the job ourselves for the amount we've paid in rates," he concluded.

It did sound pretty tough on these people, and I felt a bit ashamed of the fuss I'd kicked up over our road. At least we could get in and out in all weathers. The council seemed a bit sympathetic, too, and promised to give their road consideration.

After the delegates had withdrawn, the council discussed it with the help of a big map on the wall. This map had numerous little flags pinned on it to show what was going on around the county. It was decided that the engineer and works committee inspect the road and report back to the council. I felt it was a bit of a poor show when the group so badly needed a road. But hold on—who was going to pay for the road if it was done? Yes; all the ratepayers. The council is responsible to the whole county, and every job has to be considered in relation to other works. If they barged ahead open handed, they'd soon have the rates soaring sky-high. So they have to look before they commit themselves.

When they came to the Engineer's report the chairman looked at the clock. It was 12.25, so he proposed a luncheon adjournment before continuing. It was agreed, so they all trooped out to the hotel for their lunch.

But there was still a long way to go before the end of the business.

The Engineer's report was brief, but it gave an up-to-the-minute report on all works in progress throughout the county. There were a few questions to clear up, points not understood by some of the councillors, but it was pretty plain sailing.

One by one all the reports came forward, and were dealt with in much the same way.

At last the question of our road came up. The riding member spoke of the increased use, the extra settlers now on the road, and the wear and tear on the vehicles using it.

"A better road than the one we have now is warranted."

As he sat down, another councillor was on his feet.

"Mr Chairman, we've had a deputation here today asking for a road where there is no road. Surely, before we start giving tarsealed roads, we ought to see that everyone has metalled access. Put a few loads of metal in the holes, and clean out the ditches, and let them be thankful," he suggested.

"How much has the maintenance of the road cost us in the past year?" another wanted to know. "We might be getting to the point where sealing is cheaper than maintenance. That area has developed a lot in the last two years. We've got to look at the economics of this roading business."

Because it had been brought up before, the county clerk had information on hand to answer these questions. It was too big an expense to seal the whole road, so council decided they could afford a mile and a half of sealing and they'd keep the rest patched. Again it was a case of obtaining the most benefit from the money available, and whatever the decision it couldn't please everybody.

By afternoon tea break there were still fifteen pages to go. I thought of the work waiting at home, and left the meeting.

But members had to stay on. There was a new council representative on the college board of governors to be appointed, decisions on tenders for various works to be made, the fire brigade business to be dealt with, and the accounts to be passed for payment. I'd seen one account there for 63c. Why should council have to bother itself over a miserly 63c? But it was public money—they had to account for every cent, so there it was on the list.

Finally, I was told later, about 6 pm the meeting came to a close. Some of the councillors had a twenty-mile drive before they reached home. For their day's work they would receive a pittance that wouldn't even pay a man to do the work they would have done if they'd stayed at home. Between now and next meeting they'd have to go to more committee meetings and inspections.

And, yes—even as he put his foot inside the door, our councillor was wanted on the 'phone. A bloke up the road wanted to know how they got on about the road, and flew off the handle when he heard he wouldn't be on the sealed part. A councillor just can't win.

COUNTRY COP

Constable Murray Le Fevre

Oh to quit this wicked city, and be a country cop—nothing to do, and all day to do it. Constable Murray Le Fevre of Reefton is thirty-two, married, with three small children. This is his thirteenth year in the Police Force. He was born in Wanganui but spent most of his life on the West Coast, and he's been policeman in quiet little Reefton for nine years. In all those nine years on the Coast, our country policeman never had a day like that Friday, 24 May 1968, the day of the Inangahua Earthquake, when he went in with a small party including a Reefton doctor and a radio operator:

FROM THE TIME we got to Inangahua Landing—once we got through the cutting—there were constant rumblings, very much similar to volleys of gunfire in the hills, and the ground was shaking under our feet all the time.

I've had a little bit of experience in the Army and I've heard 25-pounders going off. This noise was very much like a battery of 25-pounders firing in the hills, a continual *woomph, woomph.* After we had to abandon the car (we ran out of road) we walked as far as the foot of the Oweka Bluff, and here was the most impressive piece of damage I've seen anywhere. A whole cliff looked as though somebody had just exploded an enormous bomb underneath it. Pieces of the bluff had gone for 200 or 300 yards, thrust completely away out on to the railway line. We had to look around to find a method of getting past this rough mess. It was far too rugged to try and cross.

We didn't know what might happen any minute with this constant rumbling and vibrations, so we went down into the riverbed instead. We made our way round the river and came out near the Inangahua Junction railway station, by the water tower. Then the first thing that struck us was the deadly quietness there—just this dead, quiet sound—except for the rumblings and vibrations. The first noticeable damage of course was to the railway wagons, which had jumped the tracks and vibrated themselves into the earth right up to the axles.

Time wasn't exactly my main concern, but I should think we reached the Junction about noon. The people living in the Junction area were all congregated on the side of the road near the hotel: very much like a picnic luncheon. They had set up trestle tables, pots with potatoes were on the boil, plenty of cups of tea were being handed out, and people were sitting round apparently in quite a good happy frame of mind. You could tell by listening to them that they were full of tension, but they were making the best of what was available at the time, thanks to Kevin Stuart, a bushman from the Junction who had got things organised.

From there we crossed over the river into the Inangahua Camp, which has the bigger population. There we found that everybody had been gathered into the Ministry of Works depot by Terry Hogue, the Ministry of Works representative in the area. He'd taken control and organised everybody there. We discussed the situation with him. The doctor who was with us, Dr Pat Hertnon, did a great job. He came back from a reconnaissance in the area to report that there was no water, no sewage, and very few houses that could accommodate the people.

So we decided there was only one answer: to evacuate everybody.

I had a Forest Service radio operator, Trevor Drower, with me, but unfortunately we couldn't get good communications back to Reefton direct, so we worked through Golden Downs Forestry, in Nelson. We could get them very clearly. We relayed our message to Golden Downs and they fed it back on the higher set to Reefton, giving out our evacuation order broken down into three phases.

The first phase was, we were going to evacuate the residents of the Inangahua area, moving out the elderly, and the women with young children up to twelve years of age. The second phase was to establish a jeep head at Rotokohu railway station, since we knew that we could drive as far as this because we'd done it earlier on. The third part was for Reefton to contact Mrs Hardy and ask her to establish a reception centre for these people coming in. At this stage we didn't know how many would be able to get out that night.

In any case I don't know what we would have done if we hadn't had helicopters. I'd like to see a helicopter in every town of any size purely for emergency works because they're so versatile. As for this evacuation, they were about the only thing possible with the whole town completely isolated by the slips. We could have got a percentage out overland, but I don't know how we'd have fared with the elderly and the very young. I doubt if we could have got them out without having a road or track opened up, and this would have taken precious time. Under the conditions, perhaps, some of these people might not have been able to stand another night. As it was, by about seven

o'clock at night, we had a good three-quarters of the people out; in fact, everybody from the Camp area except the outlying farmers.

In all, I think we evacuated 246 people from the area, in about twelve hours, allowing a lapse between seven at night and the first chopper returning as day was breaking the next morning.

Only nine of us spent the Friday night at the Junction: myself and three other policemen (who'd come in by helicopter) from Greymouth; Mr Hogue from the Ministry of Works; two teachers; and a skeleton staff left by the Hydro to keep the sub-station operating, because the sub-station at the Junction carries a main power line through to the Westport area.

We all dossed down in the best of the three classrooms that survived the earthquake. We were very fortunate. People had thrown all their mattresses and bedding and so on into the place for us, and actually I had a rather comfortable night as far as bedding went, because we had about three mattresses each underneath us, plus oodles of bedding. But as for sleep, I'd say none of us got very much.

From the Police angle, we broke up into two-men teams, operating one hour on, one hour off, right throughout the night. Our main object was to maintain security on the Post Office: money was still there although it was in the safe, and there were still remittances and stamps and all those sort of things. The hotel of course had been severely damaged and abandoned, but money was still lying all over the place on the floor and in the tills. We collected most of this up that night, and kept a general check on every house. We were patrolling the area, keeping a watch on it all through the night. Then the Hydro people of course had to check their equipment on the hour every hour.

We worked actually as three-men teams: two Police with one civilian, and the civilian would remain at the school all the time. The idea was that, should there be a further big 'quake there would be one bloke present to get the others out, to rouse them, give them warning to clear the building if anything went wrong.

Of course the room we were in shifted off its piles and jumped and bounced, and we had to clear away books and anything fitted to the walls. We took all this stuff down early in the piece, but still things kept tumbling round our ears throughout the night.

Actually the first impression, entering Inangahua that day, was very much like a lot of these films you see of a war-stricken town—everything was down in heaps all over the place, bicycles just abandoned on the side of the road, cars just left where they were, poultry and animals wandering at large, and this god-almighty silence, with the rumblings in the background all the time.

As for the people, they seemed remarkable. When the time came to evacuate, quite a lot of them had never been in a helicopter in their lives and were naturally very chary about getting into one. But they all went. Nobody presented any problem from our point of view. I would say that the people stood up to it exceptionally well, although no doubt they were apprehensive underneath.

I know I was very scared right throughout the whole thing. Fortunately I had plenty to occupy my mind so I didn't have time to dwell upon my own fears. Then I felt that it was my position not to show my own personal feelings, because people look to you for support, and if you start showing fear, people become very alarmed because they think: "Well if this bloke's scared, there's something badly wrong." Then they start having fears.

Certainly I was scared, and I'm still scared of earthquakes. I don't think anybody who's been through one can say they aren't scared of them.

Of course Civil Defence had been a hobbyhorse of mine long before this, and I think I've driven just about everybody in the town mad. It always seemed to me that there should be some system for coping with emergencies, particularly in a small place like Reefton where we can so easily be cut off from the outside world. I've always felt that we should have something within the town so that if anything did arise, we could cope ourselves until we got some help from outside. I realise I'd driven a lot of people mad by harping on this, and it made me seem a bit of an alarmist. But thank goodness we did have some form of system worked out beforehand. It certainly proved itself in the long run. We should be able to improve on it; we've certainly learnt a lot of things from this.

Every town should have an Emergency Scheme. This is definitely a must—it may never happen to you personally, but if you're prepared you can always go to the help of somebody suffering or in trouble.

The other thing is: *have the right men in the right jobs.* This is one of the beauties of our own organisation in Reefton. We've got no dissension whatsoever: everybody accepts his position, he does his best, he's prepared to learn, he's prepared to listen to others. The big thing is that everybody who is selected for a key position makes himself completely familiar with his duties, because you get unlimited helpers but your helpers are no good unless you've got somebody to direct them.

I say to anybody who is approached to take part in any Civil Defence organisation: "Don't take it on if you're not prepared to give your best. You simply must be prepared to go into it wholeheartedly, because it is for the community."

NOTE: The earthquake reached Magnitude 7.0 on the Richter Scale. The maximum-felt intensity close to Inangahua was Modified Mercalli X. A landslide killed a woman immediately, her mother died later from injuries and shock, and accidents killed three men—hitting a bridge, and a helicopter crashing. About twelve people were treated at Reefton Hospital for cuts and abrasions, but scores of people, mainly women and children, were treated daily for nervous disorders. Inangahua area has a total population of 300: all were evacuated within 24 hours. The Earthquake and War Damage Commission paid out around Inangahua $187,969 on 169 claims for damaged buildings (90) and contents. The Inangahua Disaster Fund Committee received applications for $50,000 and paid out $19,400—the total amount of the fund. Most of the settlers went back and set to work again.

Since 1848, 287 New Zealanders have died through earthquakes, 17 perishing in the Murchison Earthquake (17-6-29), and 255 in the Hawke's Bay Earthquake (3-2-31). Roughly, the risk of earthquakes in New Zealand is considerably less than in Japan, but more than in California.

INSIDE INANGAHUA

Terry Hogue

When the Inangahua Earthquake broke at 5.24 am on Friday 24 May 1968, the full burden of responsibility fell at first on the shoulders of Terry Hogue, aged 28, officer in charge of the Ministry of Works, Inangahua, Buller Gorge.

A VIOLENT UPTHRUST woke me on this Friday morning as the house seemingly took off, then settled into a violent shaking movement. There was a confusion of noise which I realised afterwards must have been the ground noises from the earthquake, the crashing of furniture and crockery within the house, and the hills disintegrating around about.

Having experienced other earthquakes, though not of this intensity of course, I thought this was the end. I thought that my wife as a mature person could face death on her own and my first thoughts were to get to my son.

How I got there I don't know now, but when I arrived in his bedroom he was sitting up in bed saying: "Daddy, daddy, the house is falling down."

I picked him up when I had managed to catch up with the bed and its mad scurrying around the room, and got him back down to our bedroom. By now the initial intensity of the shock was decreasing. Of course I was a little shattered like everything else, and after getting him into bed and reassuring my wife as best as I could in the circumstances, I started to get dressed in the dark. When the earthquake itself had died down I knew that we were all right, but I thought of other people and their surroundings: they might not be so well off.

So I immediately called out to two of our employees who live on either side of us. They answered, reassuring me that they were all right. We got their families into our bedroom and another single workman of the Ministry of Works arrived, apologising profusely for being late. He hadn't been able to find his boots in the dark. We immediately set out to search Inangahua camp. Within perhaps ten minutes or a quarter of an hour we had established that there was nobody seriously hurt, that the destruction was quite severe in some houses and less in others. People naturally gathered, I think, where they saw other people moving around.

By this time I'd put on the lights in my truck and switched on the radio to the Ministry of Works network, just in case anybody was up and about at that hour.

As the people arrived and their feelings started to become apparent, we realised that something would have to be done. We got some of the younger boys and older men to build fires in the Ministry of Works yard. We made a fireplace—in different circumstances a barbecue—out of our concrete crib-logging. We started a fire, people salvaged what they could in the half-light, and gradually drew in to this area. We told everyone else to join us where we had set up a relief centre to the best of our ability, and to bring what they could in the way of food, cooking utensils, and so on.

They arrived, some half-dressed, some slightly shocked, with what possessions they could bring at such short notice; what they could find in the confusion of their houses.

I think we had most people fed by daylight. Another community kitchen had been set up at the State Hydro camp, and then I felt that perhaps it was my duty, as officer in charge of the Ministry of Works in this area, to start organising search parties, which we did. We had our first parties out at daylight trying to get to the lower Buller Gorge area where we knew people were living. Having heard the terrible noise of the hills disintegrating, falling down, and this splitting of timber, we knew that things—well, we just didn't know what to expect there really.

Our first MOW truck went out with three men. They got only 100 yards from the camp before they were blocked by the displacement in the roading over two or three large culverts. From there, everything was on foot. We tried to get to the farming area north of Inangahua, but we were blocked again by a large slip. However, later in the morning one of the farmers walked out to say that at least they were all right. He couldn't speak of others further north.

Our first report of the injured and missing people in the lower Buller Gorge came about 9.30 with the arrival of Geoffrey Meadows, a farmer from there, who had walked about five miles out in those terrible conditions. Just after he appeared one of my own men arrived back with a similar story. I checked in the area and could find only one trained nurse, who immediately offered her services. Her husband went with her to offer what aid they could to the people in distress in the area.

In between all these things I had been putting out calls on the radio installed in my truck. I managed to raise a Gisborne station, about

9.30, I think. They answered my call and relayed a message to Westport. I could hear Westport working them back, but Westport couldn't hear me because of the atmospheric conditions. I asked for helicopters immediately to evacuate the injured to Reefton. We did not know what the outside world was like. We did not realise that we had been near the epicentre of this earthquake.

As it happened the State Hydro in Nelson had already sent a helicopter with three of their distribution engineers aboard. This pilot I believe was given instructions to make himself available to us in Inangahua when he arrived, which he did. He was John Reid.

John was immediately sent to the site of the tragedy in the lower gorge. He picked up the nurse and the two injured people and took them to Reefton, where I had told him to bring back a policeman and to supervise the finding of the missing person. As luck would have it four policemen were waiting near where he landed, wondering how they could get to Inangahua. They immediately jumped aboard the craft and soon reached us. One of these set up a radio station to try and contact a similar radio set in Reefton. Another policeman went to the site of the tragedy to supervise things there, and the sergeant from Greymouth stayed with us in the camp.

We knew by this time that there had been some damage suffered in Westport and Reefton, but we didn't know how much. So we thought it best to stay the night in Inangahua. We didn't know whether we were bound to evacuate everybody. We had no power, water, or sewerage and seventy percent of the houses were uninhabitable because of damaged roofs and damage inside.

One of the local residents was given the job of billeting everybody in the more sound houses, which he did. Some of the women took over the catering and we had a light lunch, feeling it would be better to have our main meal at night so that at least we could go to a somewhat uneasy sleep with relatively full stomachs.

While all this was happening there was a certain amount of shock apparent. The earthquake had shattered not only physical things but mental things of life too. The children, contrary to some press reports that were given, behaved, I thought admirably; there was no clinging to their parents. They played quite happily in the Ministry of Works yard which, under normal conditions, they wouldn't have been allowed to do. They played together. Some children of course knew one another, others just joined in the little parties and played quite happily there. The older people of the community behaved in a wonderful way. Because of their limitations, such as physical conditions or age, they were not as agile or able to fend for themselves as younger people could. Throughout the Friday until the time they were evacuated, they hovered almost on the fringe of the happenings in Inangahua, finding strength perhaps in the community group. But at no time were they demanding or hindering in any of their actions. They were there and accounted for, and I think they drew strength from the other people around.

Later in the afternoon we had a consultation, a meeting with Constable Le Fevre and Sergeant Peter Gruby of the Police Force, Doctor Pat Hertnon of Reefton, and Phil Meltzer, the helicopter pilot who had flown in under such remarkable circumstances. We decided to try at least to evacuate the women and children and elderly that day. So I sent out a radio message to this effect, and to ask for more helicopters, larger ones with a greater seating capacity. But shortly afterwards we received instructions from Civil Defence Headquarters, Wellington, to evacuate everybody from the area because of possible flooding in the Buller.

We had been keeping an eye on this sudden dam (made by the earthquake above Newman's Lookout) during the day as the helicopters shuttled backwards and forwards to Murchison. When the waters built up they seemed to cut their way through the dam and so we hadn't considered this a potential danger, immediately anyway. We had decided to evacuate people not for fear of flooding, but because we could not provide them with essential services. However we had now been given instructions by Civil Defence Headquarters and told that there were two large helicopters on the way.

So we got ourselves organised for this. We set up our landing strip in the Electricity Department grounds where there was a large clear lawn. We had arranged for buses to come from Reefton to the Rotokohu railway station where people would be landed to cut down the flying time of the helicopters by about eighteen minutes for each return trip; it was imperative to keep their time in the air to a minimum.

The two smaller helicopters were carrying a two adult-equivalent and the larger jet-ranger a four adult-equivalent, so all together we estimated that we could get 150 people out before dark. We ourselves were flying out by these three choppers first the elderly and the young families.

We were expecting all the time the promised larger helicopters. When they arrived [about 5.30 pm—Ed.] they immediately started to take ten or twelve adult-equivalent loads, which speeded up the evacuation considerably. We had by this time also checked the outlying communities that we could not reach on foot and found that people there were all right. We had prepared food parcels to be sent to them if we could not get them out.

There was one group of people, mainly farmers, who stayed in, in the lower Buller Gorge area, although we got most of their dependents out that night. The last helicopter flights for the evacuation were actually landing and taking off in the dark, and the last people from the lower Buller Gorge arrived in Inangahua about half-past six, or a quarter to seven. We gave them a cup of tea, if I remember rightly, and we despatched them by Land Rover through a back road. They had to clamber 300 or 400 yards over a large slip to Land Rovers, waiting on the other side, which drove them to Rotokohu railway station.

There were nine of us who stayed in Inangahua camp that night; four policemen, two of the local schoolteachers, two State Hydro employees and myself. We slept—or rather spent the night—in one of the school buildings. We maintained an all-night vigil on the off chance that somebody might still turn up: some traveller or shooter or camper who had been in the area unbeknown to us. The police went out on hourly patrols in pairs, covering both Inangahua Junction and Inangahua camp. The rest of us helped on these patrols. We checked the State Hydro sub-station to make sure it was still functioning. Others were boiling the billy to provide refreshments, making notes of the day's proceedings, and planning for the next day. We found those twelve hours that could have been quite a terrifying experience passed quite quickly.

We had plenty of company too. The local dogs that would normally snap at your heels as you went past were wandering quite closely behind you, no matter where you went. They too seemed to be seeking a little human company and reassurance. Over the next two days these dogs were probaby the best-fed in the world. They were living on the meat-stuffs that were salvaged from the deepfreezes; it was not an uncommon sight to see dogs hurrying to and fro with pork chops or legs of mutton, even fowls and ducks; taking them off to be consumed at leisure after they had thawed out a little.

At five o'clock on the Saturday morning, Fred Stacy of the NZED and I checked the sub-station. We sent a report to the Waimangaroa sub-station which would then pass it on to the Greymouth police station. We gathered up some metal containers, some oil and diesel, and set these out in the school grounds for the helicopters when they came in. We had expected them to drop in as soon as it was daylight. They were a little later than this because of fog. What a welcome sound to hear them coming up the Inangahua Valley with their peculiar *put-putting* noise, and then to see them drop in.

The strength I found during that day and those following after the quake was perhaps something that came from outside myself, something spiritual, although I didn't recognise it at the time: a deep-down assurance that there was something else other than myself working in this too.

This period has changed my life. I feel a different person: as though I do know now what I am capable of, a knowledge I couldn't have found in our normal society or way of life, so fettered in our modern society and the demands that it makes upon us.

After this, money and to a lesser extent, television, seemed completely irrelevant to the reality of living. Perhaps I'd had enough drama in all these events not to need something dished up by an institution to satisfy the inward demands for drama in normal humdrum life.

But no matter how we were compensated for our losses during this period, it's not going to replace the sentimental value of wedding presents or gifts that we had gathered over the years and other little treasures: things that had a personal significance in our life. These can never be replaced: mementos of times past in your life. It is now as though you have to start off rebuilding halfway through a life.

There are two things that I feel are very pertinent to any future Civil Defence organisation or for any emergency that might arise in this country. One is that communications between individuals, between groups, between various departments are essential. A co-ordination must be sought after and gained as soon as possible.

Then it is important to remember that Civil Defence is primarily concerned with the welfare of people in the initial stages of a disaster. Those who are working within the framework of Civil Defence or any helping organisation should bear in mind: they're dealing with a human element which has suffered perhaps severe shock, or injury. All action in the initial period after a tragedy or disaster like this should be directed wholly towards the welfare of people, individually and on a community basis.

Most of the people, I think, will remain in Inangahua and rebuild from the ruins. Perhaps it is part of man's trying to triumph over Nature that people are going to go back and pick up their broken pieces, roll up their sleeves, and get stuck in once again to the business of living.

Haerera, Haeremai

THE OLD MAN AND THE MOUNTAIN

Amelia Batistich

NO USE ALL the doctors now. Time had come. All the other times he has got up from his bed, called for his old pants and gone again to Maungariri, the mountain.

"Not time yet," the mountain say. "Some more road to go."

But not this time. He just lay there, looking at the mountain. They had much to say to each other. How good it had all been, even the bad. How much there was, even yet, to understand. Like a young man flexing his strength he had gone over his days, gathering courage for this last. If only his girls wouldn't look so sad.

He wanted to say to them: "But what do you expect? A man is ninety. That is time enough."

Time enough. Twelve years now since Mara gone. Twelve long years. Wife and workmate she had always been to him. Someone always there to talk to, as to your own thoughts. Two hearts, one heartbeat, that is married life. One heart gone, the one is only half alive.

But for his girls there is still the old smile. They look after their old dad, do all they can for him. But go and live with any one of them, that he cannot do.

"This is my place," he tells to them. "I belong to it like it belong to me."

And now they look at their old father, such grieving in their eyes, come one after another to care for him, bring the little children so the old man will not be lonely in his going. Good, good girls, they do not send him to die in stranger place. His eyes thank them for their love, speaking to each face in turn, and if another face comes up from memory it does not make them any less.

Only their mother had taken that grief to eat in her beside the older one, the first son. But another son was given, and they see him grow tall and strong like kauri sapling, till comes the war and he goes off with all the others—and one day there is telegram.

You look at paper in your hand that brings you this. It should be knife to cut into your heart. You bleed inside for your own hurt, but the scar stays hidden. Got to go on living. What else left to do?

But soon now you can lay it all down. Comes rest. Priest been yesterday, give the holy bread. Make the sign of the cross, Latin prayers saying over you. All you can say: "Amen." After, the family kneeling round the bed, the little children held up to kiss you, Mara's great-grandchildren. . . .

His mind went searching after their names to take with him to Mara. But strength was giving out. Sleep was all he wanted now.

But he had wakened. His eyes had found them all and gathered them in close. He could see, dimly, that it was not yet night but there were shadows in the room. The mountain moved back into his mind. His eyes went searching for its bulk. But someone had pulled down the blinds.

Hands fluttered. He moved, restlessly.

"He's asking for something," a voice spoke from the shadow.

Someone pillowed his head in strong arms and put a glass of water to his lips. He shook his head, somehow raised a hand to the window. It fell back on the counterpane, but they had understood. The blind rattled softly, and Maungariri the mountain came back into the room.

"Eh, old friend," the mountain said. "I been waiting."

The mountain and the man had known each other a long time. When he had first come to the place Maungariri had been wooded with low scrub, like the land. The giants had long fallen to their hollowed beds. His eyes had learned to search them out, the buried kings, and he had robbed them well. That was where the good gum lay. But you had to go down deep to find it. Stand in water all day long, shut off from the sun, look up from the darkness to the small patch of sky. The older men might complain of aching backs, stretching out their cramped limbs when day's work is done, asking if God himself knew this cursed country, swearing to get the hell out of it, soon as enough money in the tin. Get back home to wife and kids and your own place, and tell them there: Hell is in New Zealand, in places called Black Swamp and Russian Camp. But he was here to stay.

And when his own tin was filled with money he'd taken it to the store and ordered a suit from Auckland, best suit you could buy, and had his picture taken wearing it, and sent it home to his own place..

See, it is good here in New Zealand, the picture say. He still had the suit when he had the second picture taken. This time it brought him back a wife.

By that time he and the mountain had met.

"Break your back for nothing," the stock agent had told him when he bought the place. "I know that land."

"You see, mister," he'd replied to that, counting out the money for the horse. "You see."

And he'd shown them—he and Mara, toiling by him like a man. This is how you make a farm. The children came, and the little fenced-off grave. Maybe he'd felt he had more of a stake in the country with the boy there. He'd talked to the mountain about that, too.

"Don't you worry," the mountain said. "I got a long eye. I watch the child."

The mountain had had sons, too. Gone, all gone. The white man come, the brown go, leaving only his footsteps in the fern.

"Man, you only a baby yet," the mountain said. "Wait till you a million years old."

"Not got time to wait," the man had answered, and slashed and dug his way to the mountain's flanks.

"Aue!" the mountain said. "You, too."

In those young years he had not once thought about that other mountain. Been too busy in the day, too tired in the night. Only that day when telegram come, he remembered. And it was Mara, going back to all that she had left, who called it back again.

The other mountain was called Viter, for the wind. Legend and mountain and pain and suffering, it called back its child. And in the days that followed he learned to want its own hard solace. In the shadow of that mountain grew men like the rock itself, as patient and enduring. He began to call in question all that he had worked for. What is a man's hope? In the end, nothing.

So, thinking back to that other mountain, he had come to remember the stories he had heard when he was a boy, in those long nights when the old men spun their tales and the old women listened, nodding. How the sun had danced across its crest and struck the earth with fire, the morning when his grandfather died.

How the vilas wailed inside. Of battles it had seen. Of women walled into its ramparts.

He did not know what men told of Maungariri, that it had seen travail too. But once, digging near its foot, he had come upon a skull.

"Leave my bones alone!" the mountain thundered, and the storm had sent him home. But he had gone back next day and buried the skull; for this, too, had been a man.

And he and the mountain had called truce, their relationship restored, and in his own pain he had come to talk of that other mountain, laying bare the bone.

"I understand," the mountain said. "You are only stepchild of this sun."

Towards morning the man moved back from darkness. The sleepy watchers stirred. All night he had felt their hovering presences. Coming and going in waves of dream, the faces shone and faded and merged and separated again. In the conscious moments he wanted to say his last to them. But words and he were done with now.

When day came he opened his eyes and looked to the window. The sun rose red and danced across his fading vision. The world was on fire. No strength now to lift his hand.

And then the mountain spoke.

"Time now, man," the mountain said. "Go, go to your parents, to your ancestors. Go, then. Farewell! Farewell!

> *"Anei too taima haere. Me haere koe ki oo maatua,*
> *ki oo tupuna. Haere! Haere! Haere!"*

GHOSTS IN THE PLAYGROUND

Jim Borrows

THE HEAVY-EYED SHEEP grazing on the hillside were too tame and too well-fed even to move as I approached. It was spring and the flax by the creek was in full bloom. But I didn't stop to suck the peppery-sweet nectar from the red-brown flowers as I had last time I passed that way—that's something I must do again before I die.

It was still there, the steep-roofed little box that had been school; austere but kind on the windy hill top. The old sentinel pines still guarded the road fence. They had stood between the school and the mad south-westerly winds for more than half a century. One who remembered that far-off Arbor Day told me they were planted by a woman teacher who had been depressed by the treeless hilltop.

"Trees are like men," she had told her pupils as they set the young pines in the unpromising yellow clay. "They grow according to the way they're planted and the nourishment they take from the soil of life."

But there was nothing masculine about those trees. They looked like the witches and wicked stepmothers in our story books; gaunt old girls, their scrawny arms fiercely admonishing the wind. Who would dare climb them?

I slipped the iron hook out of its staple and the gate drifted open. Even the sustained squeak struck a well-remembered chord.

"I'll tan the next boy I catch swinging on that gate," the teacher had threatened. "Those hinges'll be pulled out of the post before the year's out." But there were the hinges, rusty yet solid and unmoved, twenty years later.

There were ghosts in the playground. Gay ghosts of barefoot boys and girls. Tumbling echoes of their voices to be heard by whoever had the patience to stand still and listen. A football thumping the hard ground. The flat thud of a cricket bat belting a ball. The bold music of a playground argument. The eager, raucous cheers of a fight-watching crowd . . . There, the hollow where we swapped yarns, fought, and sometimes bore the enlightening humiliation of defeat. (One did some solid growing up in the school playground.)

We fought the Battle of Hastings in the hollow one afternoon, directed by an imaginative teacher. She met some difficulties when deciding who were to be Saxons and who Normans because everyone wanted to be on the winning side, and getting the battle to go as the history books recorded was an even bigger hurdle. But the lesson was an undoubted success and when we trooped back into the classroom the salient facts of 1066 were indelibly learnt. A few weeks later we lined up outside the school porch and signed the Treaty of Waitangi. Apart from the wranglings of a couple of dissenting chiefs all went smoothly, but the teacher apparently decided from then on to conform to orthodoxy. "Playway" teaching methods were not accepted practice until twenty years later.

She was a game one, that particular teacher. The big blokes in Standard Six seemed almost twice her size, but she had them tamed from the day she arrived at the school. She had a glare and a roar—when she'd a mind to use them—that quickly shrivelled the first ill-advised buds of rebellion. On Friday afternoons she frequently walked fifteen miles to the coast to spend the weekend with friends.

The thorny acacia hedge surrounding what was once the school garden had become a veritable jungle. It had always tried to control the square it enclosed. Its questing roots robbed the ground of what little fertility it possessed, and anything we succeeded in growing was invariably stunted. But now a pink perennial sweet pea romped among the weeds. Here and there arum lillies waved exultant white banners. There were geraniums too, and some of those flowers—red with white throats—that look like miniature gladioli. All looked far healthier in their wild freedom than they ever did when we tried to cultivate them.

The hedge once provided a handy cover for all sorts of nefarious activities, like smoking lengths of cane that had been filched during handwork classes. Oh, the coughing and gasping and the pretence that we were enjoying it!

We found a sinister use for the two-inch long acacia needles. With a blob of clay placed near the sharp tip you could improvise an evil little dart with some luckless person's behind as the usual target.

There were the sad little dunnies with bracken growing high around them. The boys' tilted askew by a pine tree that had grown up beside it. The girls'—a respectful distance away—with a rambler rose twining protectively around it. Nothing remained of the shelter shed where we ate our lunches or waited for playtime showers to pass. The concrete path that once connected it with the school had disappeared under the tall paspalum grass.

There was the flagpole between the two tall windows. Frayed remnants of grey rope waved limply from its summit. The last time a flag adorned it was on the coronation day of George the Sixth. To honour the event we all had to go up to the school especially to salute the flag and sing the National Anthem. That duty completed we were sent home with a paper Union Jack and a bag of lollies.

I climbed up on the wooden box that supported the flagpole and risked a broken neck trying to peer through the window. The little cast-iron stove had disappeared from the middle of the floor, but the galvanised iron square on which it had stood was still there with a length of chimney pipe suspended above it. Apart from a table and a few wooden forms—the property of the local Women's Division group—the school room was bereft of furniture. Yet suddenly as I looked everything was as it had been: the activity; pens scratching paper; muffled whispers; stifled chuckles; the loud ticking of the wall clock that ruled the distance between "playtime" and "hometime". Funny how many memories can flit through your mind in the space of a few minutes; fleeting pictures, brilliantly coloured. A brief, nostalgic passing parade. . . .

When I was a little tacker I sat at my desk there by the stove. We used to lob bits of chalk into that knothole in the floor when the teacher wasn't looking. When I reached Standard Five I sat in the far corner (a status privilege) where the wall was decorated with pictures of People From Other Lands—some of them interestingly unclad.

In the intimate atmosphere of a one-teacher school we enjoyed an individuality impossible in a larger, more regimented, more disciplined institution. Learning was a process that was, somehow, more easily shared. One teacher, for instance, had a gift for making history infinitely more than a recital of dead facts with a date attached. Standard One to Standard Six used to listen with intense interest as the great canoe fleet set out from Hawaiiki, or Robert Bruce led the clans to victory at Bannockburn, or Richard the Lion Heart fought the Turks in the Holy Land.

We returned to school after one summer's holiday with a new word added to our vocabularies: consolidation. The school was to be closed and we were to attend the District High School in our county town.

Marvellous!

We'd be going to a big school.

What was more, there'd be no more walking. We'd have a bus. We were impatient for the weeks to pass and the great change to come about.

So at the end of the term we trooped into the little schoolroom for the last time. Our teacher had spent an hour or so at the blackboard preparing something to honour the occasion. He'd drawn a couple of children having a good old weep and beside it he had written about the school; the things we had done there, and the things we were going to miss. The small, important things we'd learned about life and about living. The lessons we'd remembered and those we'd forgotten. The games. The songs. The laughter . . . . How I wish I could remember exactly what was written there in green, red, and yellow chalk, for the words fitted the occasion perfectly. At the time we thought the gesture rather sloppy, but before the year was out we were to know that every word was true.

It was still there all these years later. It could have been written yesterday. The colours had hardly faded. I suppose it was significant that the groups that met in the little building had been careful to preserve the epitaph—and not only because most of them were ex-pupils. Some well-meaning vandal had added Sir Walter Scott's bit about "Breathes there the man, with soul so dead . . ." and missed the theme by a mile.

The little school underwent some renovations and had some additions made to it a year or two ago to make it more suitable for its present function of district meeting-place. Its appearance had changed considerably, but at least it's being used. I have seen similar buildings in other parts of the country—unhappy derelicts left to crumble unmourned in jungles of blackberry and bracken.

But that's progress, I suppose. The days of those one and two-roomed schools have almost gone, but the place they fill in the story of our country's development is an honoured one.

OPEN COUNTRY CALLING

Arrows point to location of stories, numerals indicate the page numbers on which stories appear in the book.